MW01626031

ANTHROPOCENE

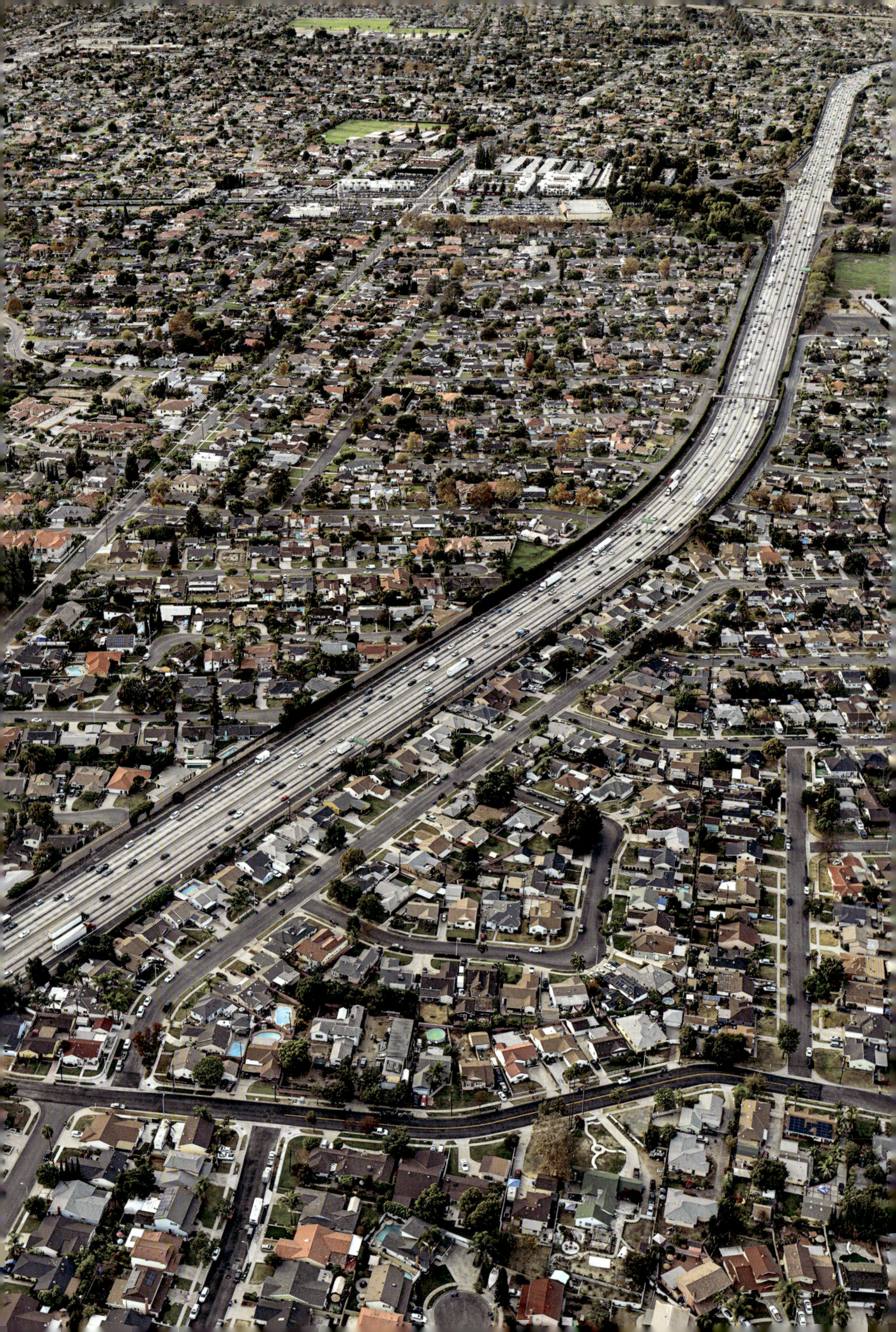

ANTHROPOCENE

BURTYNSKY BAICHWAL DE PENCIER

Published by

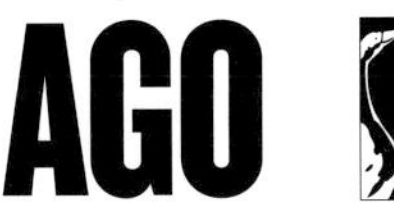

In collaboration with

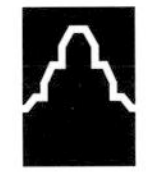

National Gallery of Canada

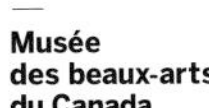

Musée des beaux-arts du Canada

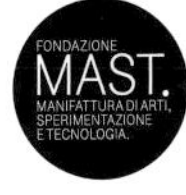

COVER
Edward Burtynsky, *Tyrone Mine #3, Silver City, New Mexico, USA* (detail), 2012.

PREVIOUS PAGE
Edward Burtynsky, *Highway #8, Santa Ana Freeway, Los Angeles, California, USA* (detail), 2017.

FOLLOWING PAGES
Camera secured to train at the Gotthard Base Tunnel in Switzerland. Courtesy of Anthropocene Films Inc., © 2018.

Nicholas de Pencier capturing the Gotthard Base Tunnel in Switzerland. Courtesy of Anthropocene Films Inc., © 2018.

10 **Foreword**
Stephan Jost, Marc Mayer, and Isabella Seràgnoli

13 **Far and Near:
New Views of the Anthropocene**
Sophie Hackett

35 **The Anthropocene and Its "Golden Spike"**
Colin Waters & Jan Zalasiewicz

45 **"How Anthropo-scenic!":
Concerns and Debates about the Age of the Human**
Karla McManus

59 **Works**

189 **Life in the Anthropocene**
Edward Burtynsky

197 **Our Embedded Signal**
Jennifer Baichwal

205 **Evidence**
Nicholas de Pencier

209 **Adams, Adams, Baltz, Burtynsky:
The Role of Landscape in
North American Photography**
Urs Stahel

221 **The Art Museum and the Anthropocene**
Andrea Kunard

231 **Notes**

243 **List of Works**

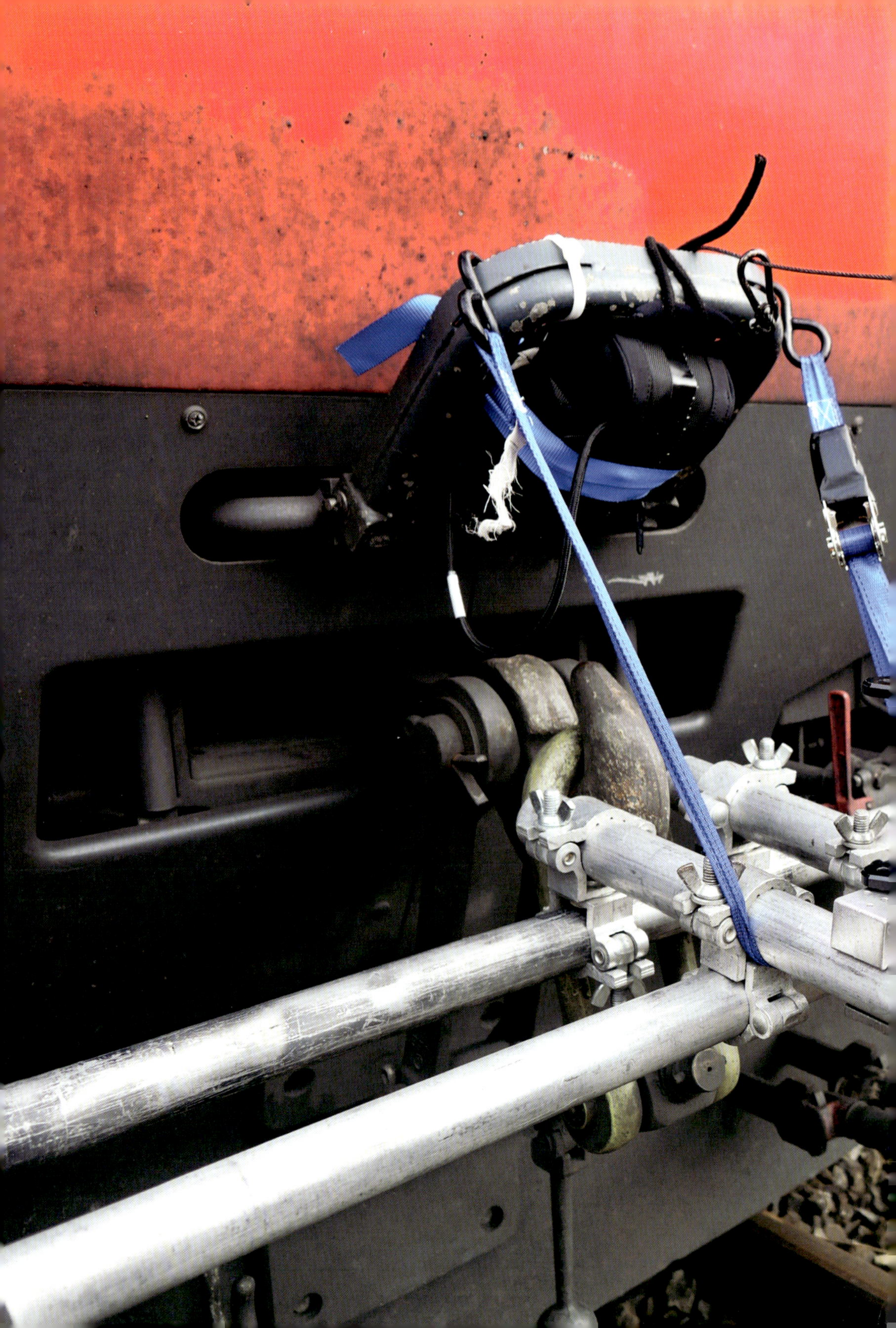

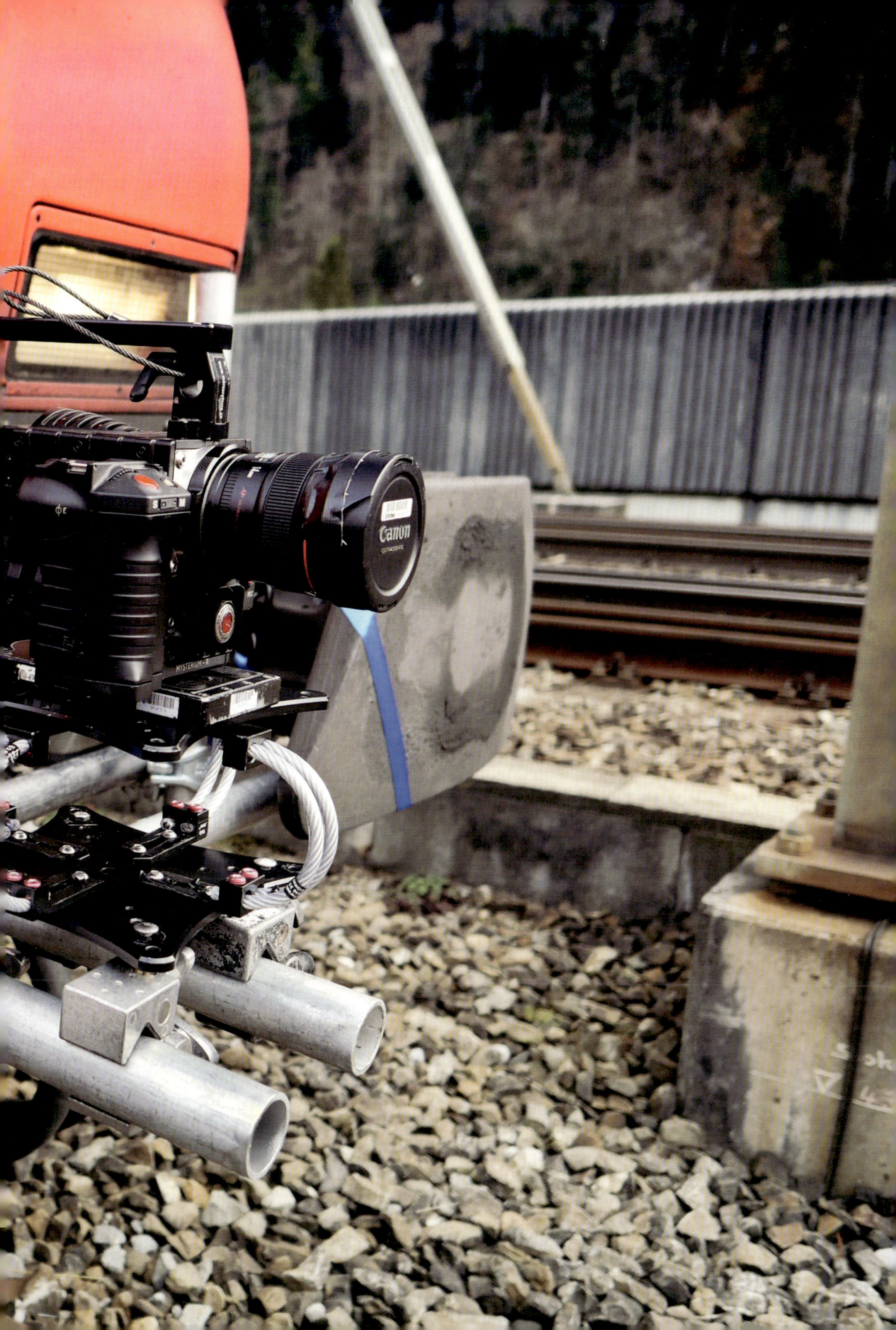
Canon
MYSTERIUM-X

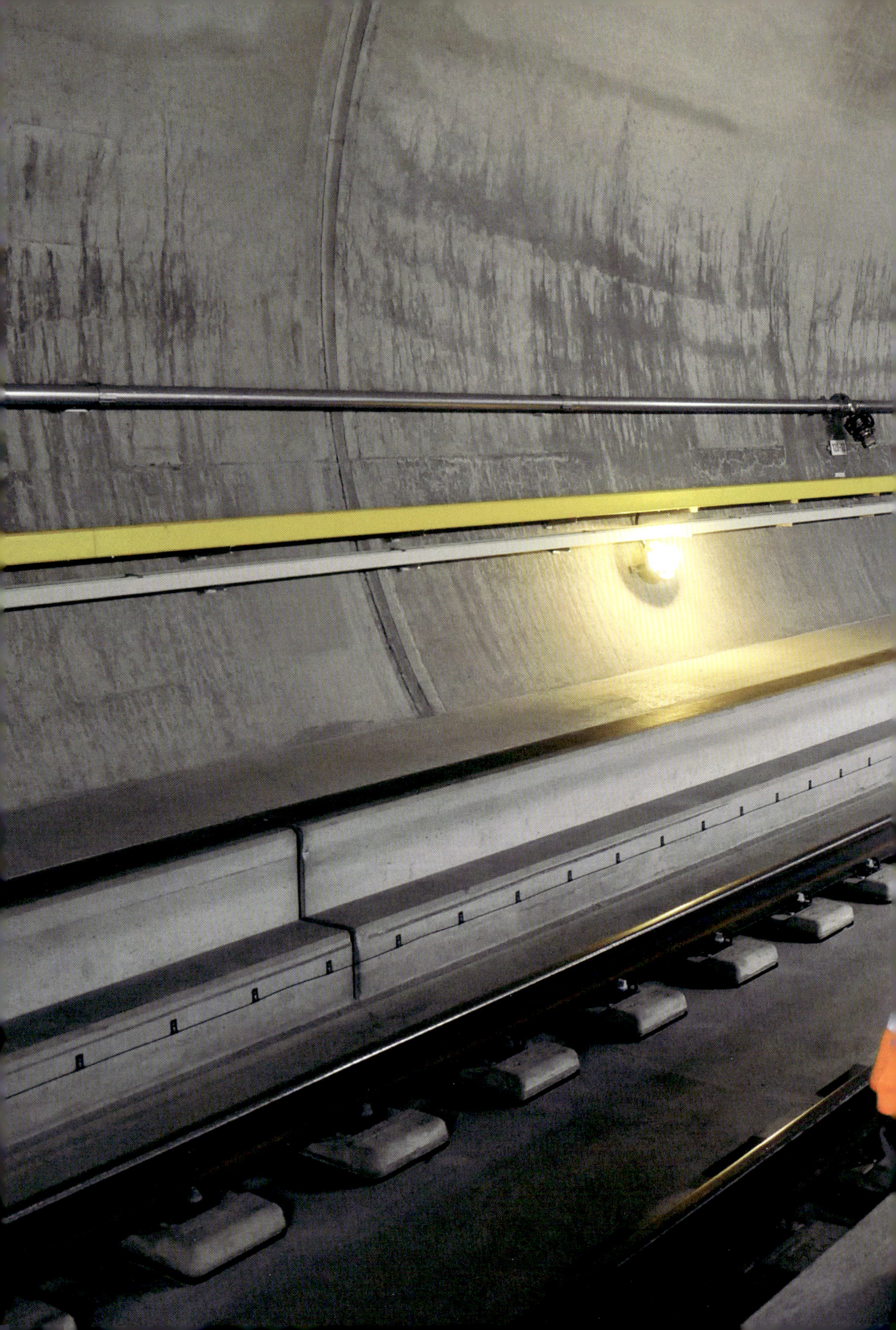

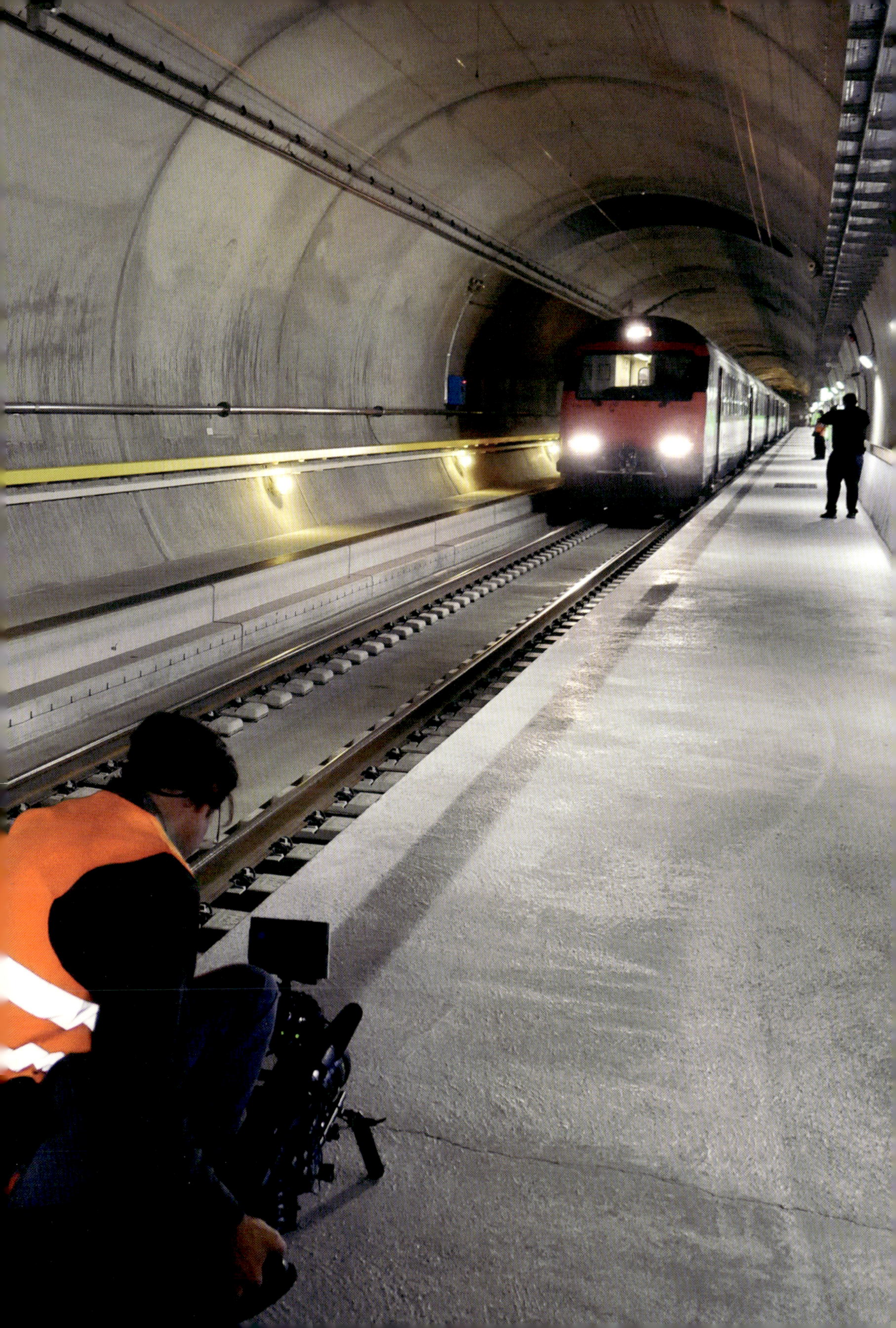

Foreword

Anthropocene is the culmination of an ambitious, four-year-long collaboration by the artists and filmmakers Edward Burtynsky, Jennifer Baichwal, and Nicholas de Pencier. Spanning photography, film, and virtual and augmented reality, this rigorously researched project invites viewers to bear witness to how the planet has been irrevocably transformed by human activity. Through stunning and innovative imagery, *Anthropocene* achieves its goal of capturing the massive scope of human effects on land, sky, and water. We have no doubt that our audiences will leave with an unshakable sense of both awe and responsibility.

We are pleased to present this important new project, the result of a fruitful collaboration between three institutions: *Anthropocene* is organized by the Art Gallery of Ontario (AGO) and the Canadian Photography Institute (CPI) of the National Gallery of Canada (NGC), in partnership with Fondazione MAST.

This marks the first time that the Art Gallery of Ontario in Toronto and the National Gallery of Canada in Ottawa will present simultaneous, complementary exhibitions. We hope that some visitors will be able to visit both cities to experience the entire scope of the project. In conjunction with the exhibitions, the feature documentary film *Anthropocene* will also premiere in the fall of 2018. The show will then travel to Fondazione MAST, Bologna, in the spring of 2019 for its European debut.

Many thanks are due to the co-curators of the exhibition, AGO Curator of Photography Sophie Hackett, CPI Associate Curator Andrea Kunard, and MAST Curator Urs Stahel, who shepherded the project through its many stages with utmost commitment. Staff across all areas of the AGO made this project possible with their dedication and experience, especially the exhibition teams led by Hillary Taylor, Project Manager; Nadia Abraham and Shiralee Hudson Hill, Interpretive Planners; and Katy Chey, Aleksandra Grzywaczewska, and Kristina Ljubanovic, Exhibition Designers.

At the NGC, Karolina Skupien, Senior Exhibitions Manager, oversaw all details of the exhibition with much professionalism. Béatrice Djahanbin, Education Officer, capably handled didactic components, and David Bosschaart, Senior Designer, expertly developed all aspects of exhibition design. A heartfelt thanks must also be extended to the proficiency of the NGC's Technical, Publications and Copyright, Multimedia, Facilities, Planning and Management, and Marketing and New Media teams. Many thanks also go to MAST staff and to all those who contributed to make the exhibition possible in Bologna.

This publication was spearheaded by the AGO's publishing department, which is led by Jim Shedden, Manager of Publishing. The contributors to this book have thoughtfully and compellingly illuminated the complex themes of the project; this includes the curators and artists, Colin Waters and Jan Zalasiewicz of the Anthropocene Working Group, and art historian Karla McManus.

We are immensely grateful for the generosity of the supporters who have made *Anthropocene* possible. A very special thank-you goes to the Art Gallery of Ontario and the National Gallery of Canada's Presenting Sponsor Scotiabank, Founding Partner of the Canadian Photography Institute at the National Gallery of Canada. Both institutions also thank TELUS for its generous partnership and contributions to the exhibition. The AGO also extends sincere thanks to its Lead Supporter, the Hal Jackman Foundation. In addition, the AGO is grateful for generous support from the following donors: Greg and Susan Guichon; Richard M. Ivey; Richard and Donna Ivey; Suzanne Ivey Cook; Rosamond Ivey; and Robin and David Young. Generous assistance is also provided by Michael Barnstijn and Louise MacCallum; The McLean Foundation; Gretchen and Donald Ross; and the Donner Canadian Foundation. The AGO also thanks its government partners: the Canada Council for the Arts and the Ontario Cultural Attractions Fund, a program of the Government of Ontario through the Ministry of Tourism, Culture, and Sport, administered by the Ontario Cultural Attractions Fund Corporation.

Edward Burtynsky, Jennifer Baichwal, and Nicholas de Pencier have shared with us an exceptional body of work that captures the entangled nature of humans and our environments. The Art Gallery of Ontario, the National Gallery of Canada, and Fondazione MAST are proud to bring *Anthropocene* to national and international audiences, and to play a role in furthering the dialogue around what it means to be alive right now, in this crucial moment on earth.

STEPHAN JOST
Michael and Sonja Koerner
Director, and CEO
Art Gallery of Ontario

MARC MAYER
Director and CEO
National Gallery of Canada

ISABELLA SERÀGNOLI
President
Fondazione MAST

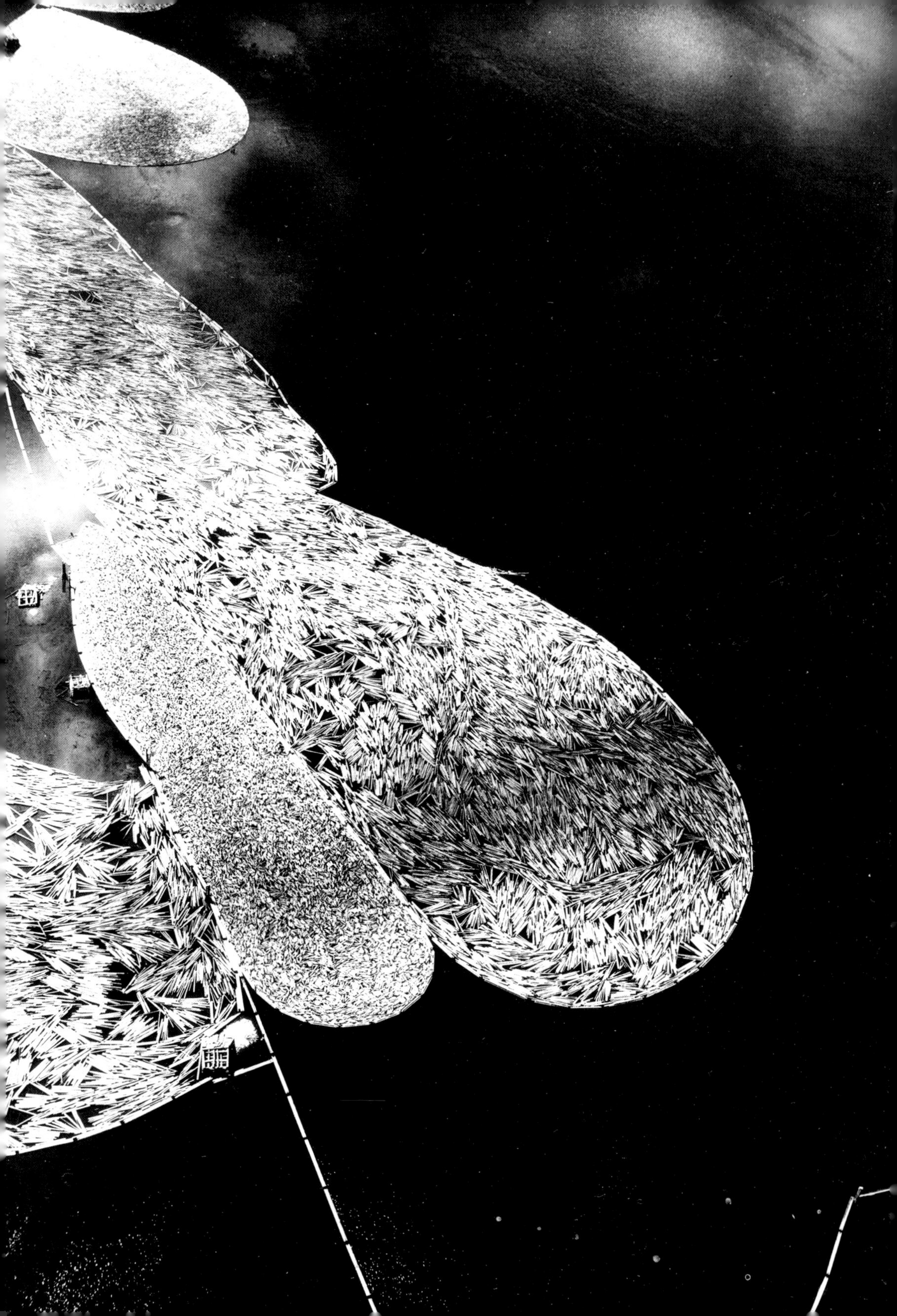

Far and Near: New Views of the Anthropocene

Sophie Hackett

Filmmaker Jennifer Baichwal remembers precisely when the idea for *Anthropocene* was sparked. In 2014, she and photographer Edward Burtynsky were in Washington, D.C., to screen their film *Watermark* (2013), made with Nicholas de Pencier. Jennifer turned to Ed and asked: "What if we could make 'anthropocene' a household word?" During the filming of *Watermark,* Baichwal had been struck by the state of the Colorado River delta, how the fingers of its tributaries emerged from the soil, reaching for a river that was no longer there, the result of an agreement between Colorado and California to divert water for crop irrigation.[1] It was stark evidence of humanity's ongoing and accelerating impact on the planet.

To begin to answer this "What if?" question, long-time collaborators Baichwal, Burtynsky, and de Pencier collectively conceived all aspects of an ambitious new project. Now known as *The Anthropocene Project,* this undertaking includes a documentary film, two publications (including this one), a national education program in partnership with the Royal Canadian Geographical Society, and a series of exhibitions. The framework for the artists' inquiry draws heavily on the research of the Anthropocene Working Group (AWG), a group of thirty-seven scientists who have been gathering evidence since 2009 to make the case that we are no longer in the Holocene—our current geological epoch—but rather in a new epoch, the Anthropocene.

OPPOSITE PAGE
Margaret Bourke-White, *Log Rafts, Canadian International Paper Company, Lake St. John, Canada* (detail), 1937. Gelatin silver print, printed later, 35.5 × 27.7 cm. Art Gallery of Ontario, purchase, 1974. © Estate of Margaret Bourke-White/SODRAC, Montréal/ VAGA, New York (2018). 74/19.

For the last five years, the trio has been driven by the question of how to give compelling aesthetic form to the evidence that has aggregated in the geological record through persistent and globally interconnected human activity. The topic links deeply to the artists' previous collaborative work, the films *Manufactured Landscapes* (2006) and *Watermark* (2013), but also to their independent work. In nearly every series in his career to date, Burtynsky has focused on the human-altered landscape. The scope and breadth of these series have increased in ambition and complexity in recent years as he has moved from specific industrial sites, such as *Railcuts* (1985) and *Quarries* (various locations, 1991–1992, 1993, 2000, 2006), to economies—for example, *China* (2004–2006)—and to networks of production and use, such as *Oil* (1999–2010) and *Water* (2009–2013). Baichwal and de Pencier, in all of their films, work from an intense engagement with place, examining how the character of a specific landscape can shape cultures and individuals, constitute a sense of self, and feed creative expression—a "dialectic of thought and geography," as one scholar has put it.[2] What unites the artists' work is a sense that the landscapes we inhabit and the landscapes that support us, though in some cases geographically distant, are not separate—and further, that the misuse and permanent alteration of these landscapes and their resources is not only a threat but presents a potential and fundamental change to our beings.

What also unites the artists is the sense that documentary images—be they photographs, films, or augmented reality installations—can reveal the extent and character of some of the most affected places in the world, locations that, as Baichwal says, "we're responsible for but would never otherwise see."[3] Guided by the AWG's research, the artists travelled to sites where our anthropogenic activity is particularly visually evident. With the central question of how to best capture and convey the sheer scale of this activity, it is no surprise that the artists have produced large-scale photographs and film installations—but they also push beyond singular experiences of a photograph or a film, the still or moving image. Driven by the desire to deliver compelling and close experiences of places particularly marked by human activity, they have combined their efforts to produce new works that amplify the qualities of their respective mediums in a series of large-scale, high-resolution murals with film extensions. They have also collaborated to explore entirely new media altogether, in augmented reality installations, gigapixel essays, and in 360° virtual reality films.[4] As de Pencier points out, "[Lens-based media] represent the most technological, most anthropogenic, of the art forms, and that is why we feel confident deploying them for this project."[5]

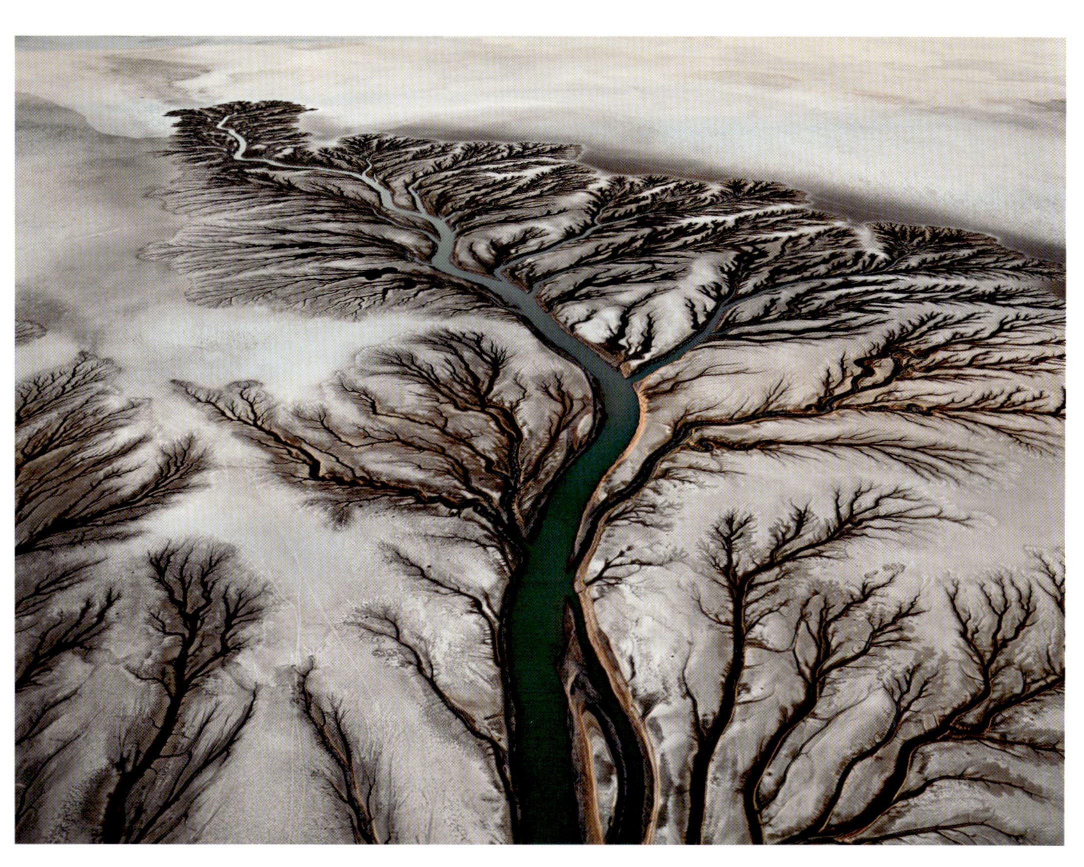

Edward Burtynsky, *Colorado River Delta #2, Near San Felipe Baja, Mexico,* 2011. Chromogenic print, 99 × 132 cm.

The View from Above

Many scholars have pointed out how our experience of humanity's impact has been predominantly visual and sensorial.[6] Charts, graphs, and photographic images are regularly deployed to give visual form to a range of phenomena: melting ice caps; the effects of rising water levels on cities; areas of deforestation. The view from above has been one predominant method for relaying these changes. This grows out of a long history, at first, of seeking out elevated vantage points and later creating views from the air, from various aircraft, spacecraft, and now also drones. The impulse to try to see ourselves in the world, apart from ourselves—a seemingly paradoxical aim—is chiefly evident in the history of cartography.[7] Long before the possibilities of flight and of photography, humans created flattened, overhead views of their surroundings. With the advent of photography in 1839, photographers regularly documented new bridges, railroads, buildings, monuments, and city skylines—testaments to human ingenuity as well as to growing colonial and commercial networks. Canada was no exception: building a railroad to link the vast territory was part of Confederation in 1867, a way for settlers to lay further claim to the landscape. The photographers followed.

Such representations found new purposes and reached new levels of ubiquity with the invention of the airplane and dry plate photography, particularly after the First World War. The usefulness and aesthetic impact of aerial views have remained closely intertwined since their first use. The potent mix of abstraction and information in these photographs continues to fascinate, as the viewer absorbs and then recognizes the information, a move from the unfamiliar to the familiar. This has meant that such views are one of the few types of photographs that have been able to operate at the nexus of many fields—industry, art, urban planning, geography, and the environmental movement, to name a few—which earns them greater persuasive power.

The aerial mapping of enemy territory throughout the First World War was key to military strategy, but the original function of such images didn't deter Beaumont Newhall, the first photography curator at New York's Museum of Modern Art, from including sequential aerial views of a bombardment in progress in the museum's first survey of the medium's history in 1937.[8] Early in her career, Margaret Bourke-White (1904–1971)—whose 1936 photograph of the Fort Peck Dam in Montana graced the first cover of the newly relaunched *LIFE* magazine that year—made aerial views to document industrial structures and operations for magazines and corporate clients; these photographs are now at home in art museum collections.[9] William Garnett (1916–2006) had a similarly expansive professional trajectory, including museum shows, magazine commissions, a teaching position with the University of California, Berkeley's prestigious College of Environmental Design from 1968 to 1984, and a collaboration with architect and environmental activist Nathaniel Owings in *The American Aesthetic*, the pair's 1969 book about land use. Ansel Adams and Nancy Newhall also included Garnett's work in *This Is the American Earth*, a book published by the Sierra Club in 1960 to raise awareness of the need to preserve nature.

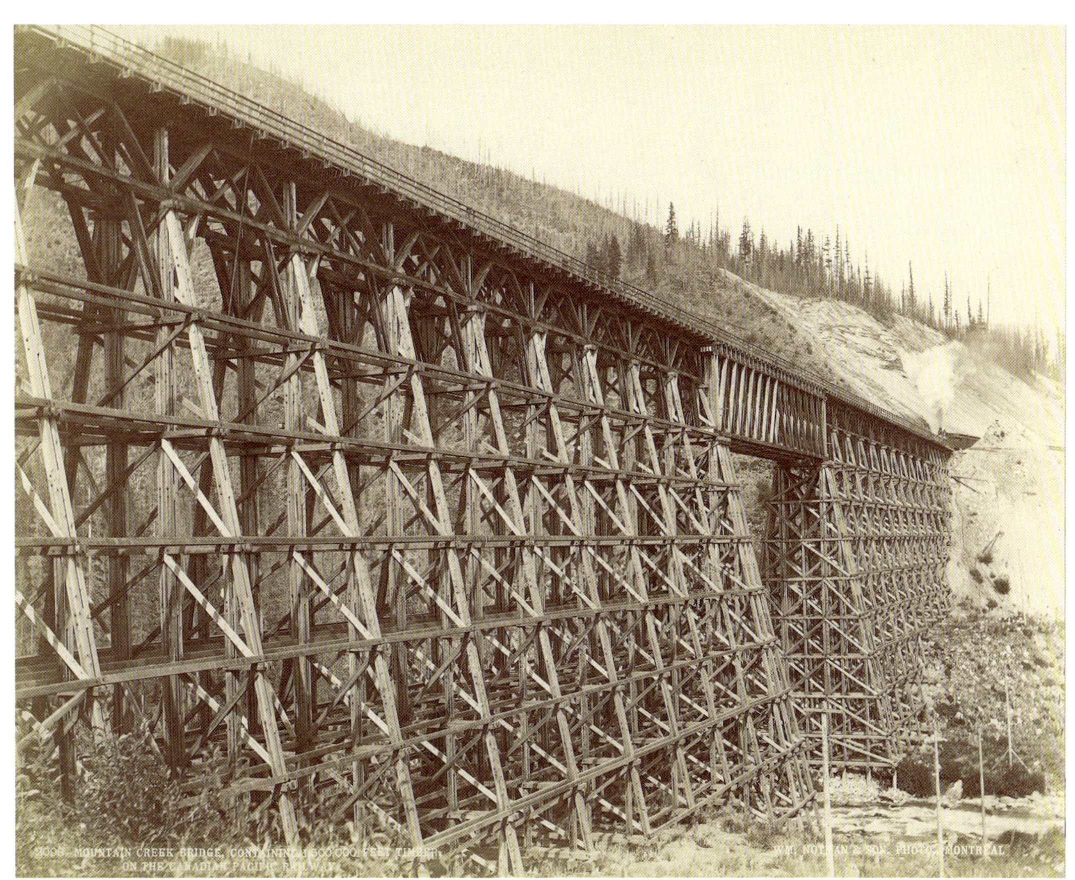

TOP
William McFarlane Notman. *Mountain Creek Bridge, on the Canadian Pacific Railway, B.C.,* 1889. Albumen print, mounted, 27 × 35 cm. Art Gallery of Ontario, gift of Lynn and Stephen Smart, 2012. 2012/144.

FOLLOWING PAGES
Pages 36 and 37 in *This Is the American Earth* by Ansel Adams and Nancy Newhall with photos by William Garnett. Published by The Sierra Club, 1960. © 1960, Nancy Newhall, © 2018, the Estate of Beaumont and Nancy Newhall. Permission to reproduce courtesy of Scheinbaum and Russek Ltd., Santa Fe, New Mexico.

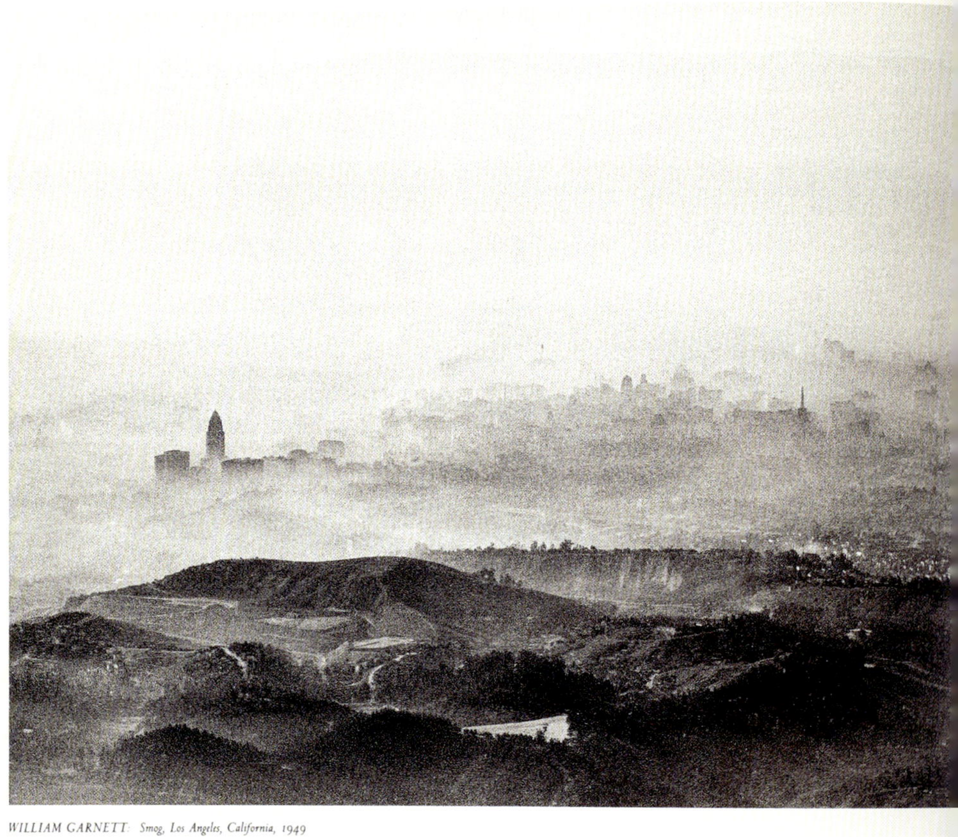

WILLIAM GARNETT: Smog, Los Angeles, California, 1949

Hell we are building here on earth.

Headlong, heedless, we rush
    —to pour into air and water poisons and pollutions until dense choking palls of smog
        lie over cities and rivers run black and foul
    —to blast down the hills, bulldoze the trees, scrape bare the fields
        to build predestined slums; until city encroaches on suburb,
        suburb on country, industry on all, and city joins city,
        jamming the shores, filling the valleys, stretching across the plains

WILLIAM GARNETT: Housing Development, Los Angeles, California
Series of four aerial photographs

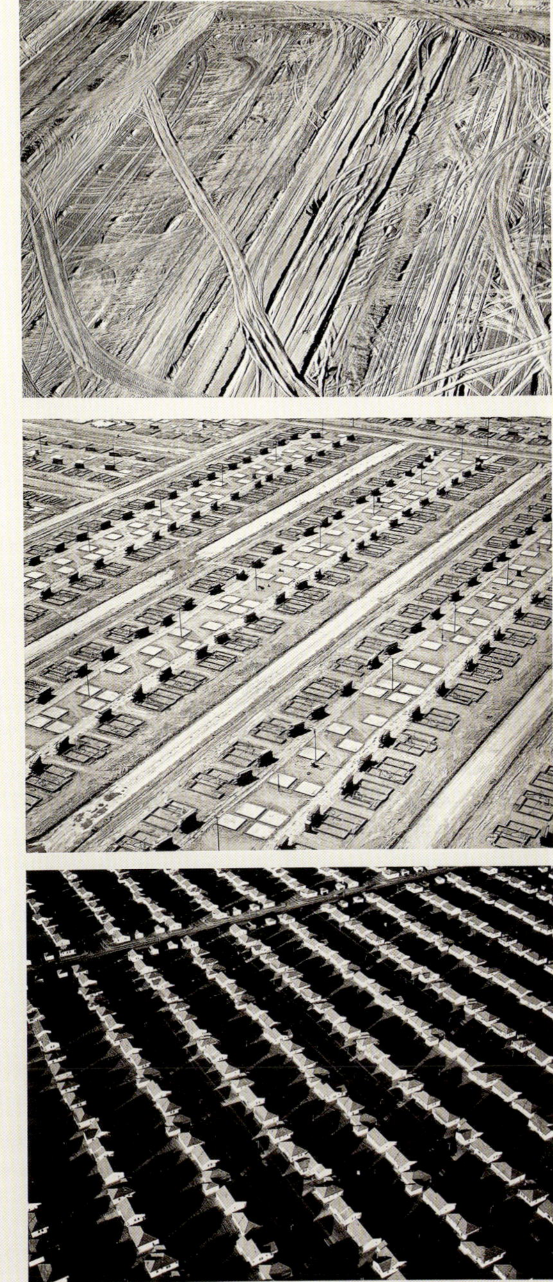

Indeed, the aesthetic impact of the aerial view has proven a useful tool, particularly with respect to the environmental movement.[10] Nancy Newhall characterized the advent of Lakewood, California, which Garnett documented throughout the community's construction process between 1950 and 1954, as "the hell we are creating here on earth."[11] In 1952, urban planner, sociologist, and human geographer E. A. Gutkind published *Our World from the Air*—an international compendium of aerial views, chiefly of the built environment. Lewis Mumford, the noted American historian, sociologist, philosopher of technology, and literary critic, provided the book's introduction, declaring:

> The earth as we know it holds both a promise of heaven and a threat of hell. If we are heedless of our environment and of human needs, the processes now at work will doom our whole planet to destruction—slow destruction through the removal of the forest cover, the erosion of the soil, the lowering of the water table, the over-exploitation of the land, or swift destruction from the pollution of the atmosphere and the stratosphere, through perverse misapplications of atomic energy to industry, warfare, and to mass genocide. The problem of our generation is to summon up enough love and fellow feeling, for both the earth and its inhabitants, to turn every habitable part of it into a permanent home.[12]

Postwar exhortations aside, Mumford supported Gutkind's sense that the aerial view would provide invaluable information and insight for improving humanity's lot—or more specifically, humanity's relationship to the environment. The connection between such views from above and the environmental movement was further cemented in 1968 when Stewart Brand included a composite colour satellite image of the earth on the cover of the first *Whole Earth Catalog,* an event that many see as a turning point in the development of environmental awareness and activism on a global scale.[13]

It is important, too, to note a particularly Canadian condition as context: the National Film Board of Canada (NFB), a government agency, was founded in 1939 under the direction of John Grierson. Through documentary films and photographs (the Still Photography Division was founded in 1941), the NFB was set up to create a visual record of Canada and its people and to awaken the highest ideals of citizenship. As Carol Payne notes,

> These photographs were often jingoistic, but they also included some of the most innovative photo-journalistic images made in this country at the time. They were consumed regularly by millions of Canadians as well as international audiences in newspapers, magazines, NFB-produced books, filmstrips, and exhibitions. In effect, the division served as the country's image bank, producing a governmentally endorsed portrait of Canadian society.[14]

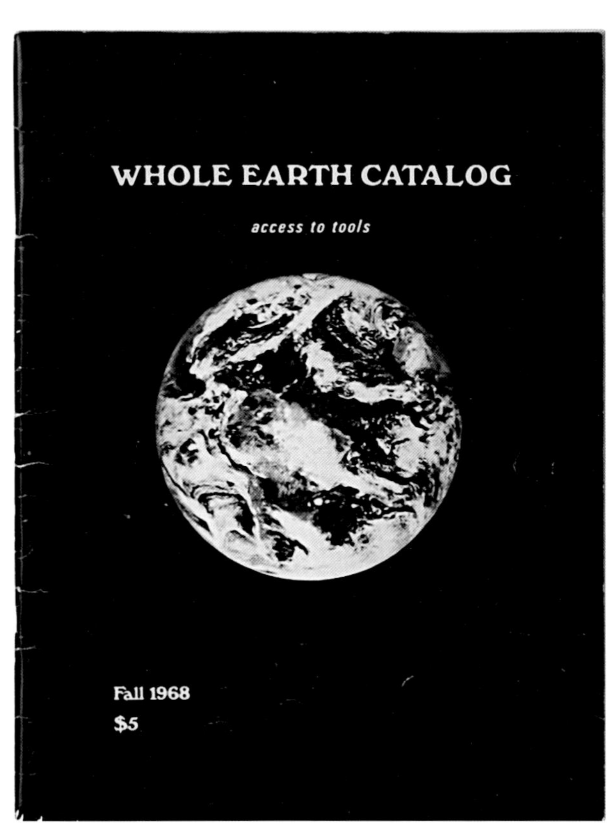

Whole Earth Catalog, Fall 1968, Issue 1010.
Edited by Stewart Brand.

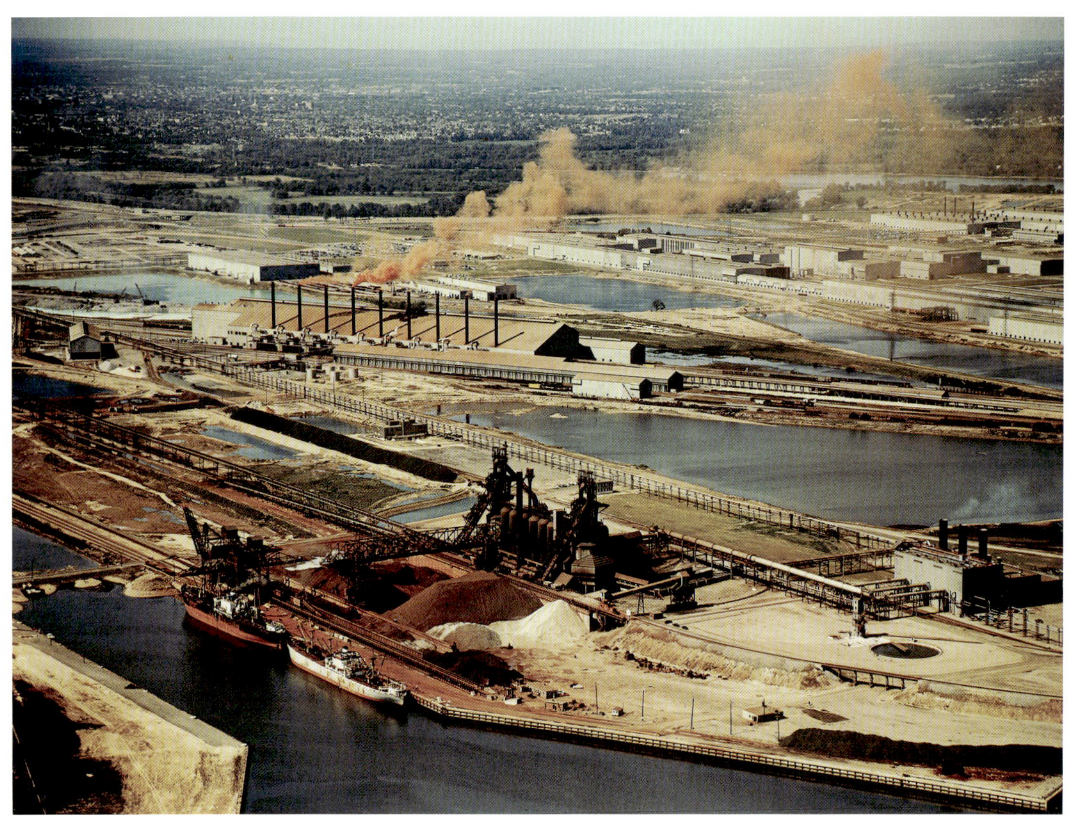

George Hunter, *Dofasco and Stelco Steel Mills, Hamilton, Ontario,* 1954. Dye transfer print, 31.2 × 42.1 cm. Art Gallery of Ontario, gift of George Hunter, R.C.A., 2010. © Canadian Heritage Photography Foundation. 2010/260.

Canada's visual culture in the twentieth century was profoundly shaped by the NFB. Not only did it provide a visual culture backdrop for Burtynsky, Baichwal, and de Pencier to work from and against, but I would also argue that the NFB provides a precedent for their collaboration across different media in *The Anthropocene Project*.

During this postwar period in Canada, the aerial view was chiefly deployed for industrial, commercial, or nationalistic purposes. For instance, in the lead-up to the centennial of Canadian Confederation in 1967, the NFB commissioned films and published books, like *Helicopter Canada* (1966) and *Canada: A Year of the Land* (1967), which highlighted the vastness and diversity of the country's landscapes and its industries.[15] One of the most prolific photographers during this time was George Hunter (1921–2013), who not only worked freelance for the NFB but also had many corporate clients.[16] His two specialties were mining photographs and aerial views. Hunter's photographs highlight a growing postwar economy based on resource extraction. He was sought after for his pioneering low-altitude aerial views in the 1950s and 1960s, works that make vivid the era's industrial operations. This would change in the 1980s, when the growing influence of the environmental movement gave new meaning to images like Hunter's view of steel mills in Hamilton from 1954: pictures that had once been understood to depict progress and productivity were now read as records of environmental degradation.[17]

This is the context in which Burtynsky, Baichwal, and de Pencier began their work. With *Manufactured Landscapes*, their first collaboration, Baichwal and de Pencier sought to visually translate Burtynsky's approach to making photographs, rather than produce a didactic portrait of an artist and his intentions. For *The Anthropocene Project*, they seek to visually translate the scientific findings of the AWG. The aerial view provides one key method, as do the immersive technologies of virtual and augmented reality (VR and AR respectively). Overall, the impact of this project rests in the aggregate experience of the varied media in the exhibition, an acknowledgement that the problem is too complex and too global for a single artist or for a single medium.

A Collaborative View

Burtynsky began to work from the air in earnest for his *Water* series and during the filming of *Watermark*. In 2009, *National Geographic* magazine commissioned Burtynsky to document water usage in California.[18] The resulting series of photographs, published in April 2010, provided him with his first sustained opportunity to create aerial views, a mode of working that is now his primary one, as it allows him to so effectively document the scale of anthropogenic activity on the surface of the planet.[19]

True to form, for *The Anthropocene Project*, Burtynsky delivers arresting, graphic, colour scenes in each of his photographs. The border between cultivated and uncultivated land at a palm plantation in Borneo is expressed as a sharp line. A bulldozer pushes chalky overburden from a Florida phosphorus mine at the edge of a tailings pond that is electric lime green. In China, a row of six moulds sits implacably opposite the hundreds of concrete tetrapods they were used to produce, stacked like massive jacks. Burtynsky works within the tradition of aerial views and brings his distinct vision to bear on each image: a compelling play of formal elements gives rise to both disorientation and discovery.

Similarly, Baichwal and de Pencier move viewers between the familiar and the unfamiliar in their films. We follow a man walking down a road in the Dandora landfill outside of Nairobi, Kenya. There are tall mounds on both sides of the road, forming a small canyon. It takes a few moments to realize that the mounds are entirely composed of plastic waste. A long, continuous, perpendicular view of oil refineries in Houston, Texas, slowly brings awareness of the extent of the refineries; it seems we will never reach the end. As the footage turns from day to night, these structures come alive in a new way: daytime details disappear and nighttime lights define the territory. The refineries remain constant, always operating, turning crude oil into a range of products.

The photographs and films are closely linked; the three artists decided on and travelled to each location together. The power of each artist's individual expression in their chosen medium becomes particularly evident in works made in the same area. For instance, in the area near Gillette, Wyoming, rich in both oil and coal, Burtynsky photographed an open-pit coal mine from above (page 128). Part of the vast terraced pit sits in the foreground, spreading towards us, dwarfing the mine buildings just beyond, and off to the right, a network of similar pits is visible. A thin horizon line beyond anchors the whole composition, contrasting the open, flat terrain with the depth and extent of the pit. In their film installation *Coal Trains, Wyoming, USA* (2018, pages 130–131), Baichwal and de Pencier keep us on the surface and focus on a rail line that transports the extracted coal. They position the camera at first on the tracks, and it remains stationary as the rail cars loaded with coal enter the frame and keep coming. The single movement the filmmakers make with the camera is to pan upwards, revealing that same thin horizon line and the same open, flat terrain of the photograph, with the rail line vanishing far off into the distance. Together, these two works deliver both vertical and horizontal perspectives, and spatial as well as time-based experiences of the ways coal mining marks the Wyoming landscape, and the intensive resources required to extract and transport this form of fuel.

OPPOSITE PAGE TOP
Jennifer Baichwal and Nicholas de Pencier, *Coal Trains, Wyoming, USA* (film still), 2018.

OPPOSITE PAGE BOTTOM
Edward Burtynsky, *Coal Mining, Near Gillette, Wyoming, USA*, 2015.

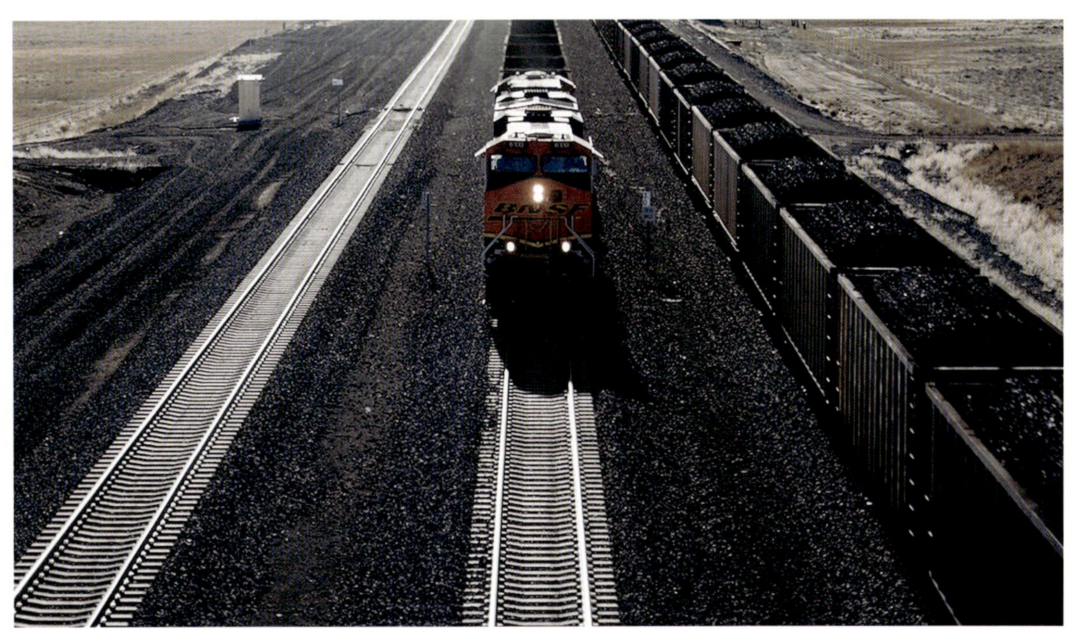
BNSF

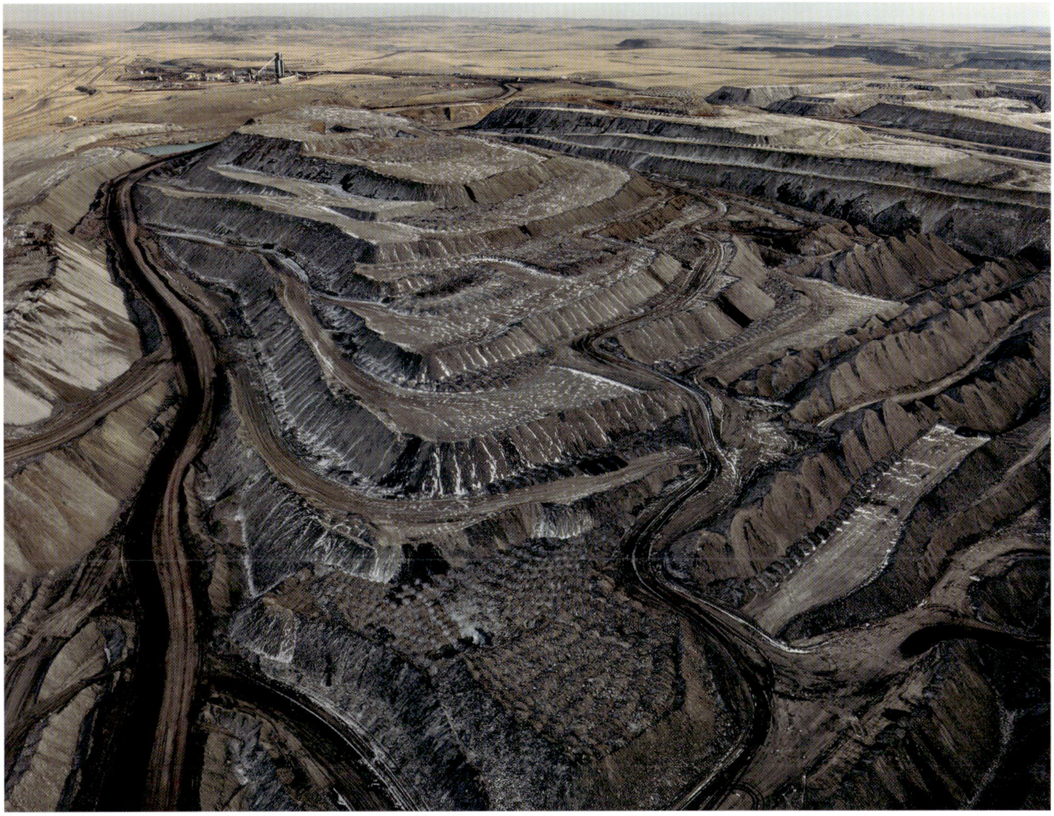

A different kind of aggregate impact becomes possible, too, as our awareness of the interconnections among the pictured locations builds. For instance, looking at a range of sites where industrial fertilizer is either produced or used—a potash mine in Russia, phosphorus tailings in Florida, the fields of the Imperial Valley in California, the greenhouses of Spain—we begin to perceive a network animated by supply and demand, connected by a commodity that never exactly comes into view. Beyond simply pairing photographs with films, a series of high-resolution murals with film extensions and a group of augmented reality installations further evince the all-encompassing yet hard-to-capture nature of the problem at hand.

The high-resolution murals in the exhibition are the largest photographs Burtynsky has created to date: they range in size from 2.4 × 7.3 metres to 3.7 × 7.3 metres. Stitched together from hundreds of individual photographs, the murals present views equally focused throughout the picture plane, and extraordinary detail that exceeds what the human eye can see. The large-scale tableau format is by now a well-established aesthetic choice for artists working with photography—works at that scale fill the field of vision and create a physical relationship with viewers, who often must move a number of steps forward, back, or side-to-side to engage with the work. Beyond this, however, the increase in scale serves a unique purpose in each mural.

The aerial view of Lagos, Nigeria (pages 75–77)—one of the world's fastest-growing cities, a place Burtynsky has described as the "hyper-crucible of globalism"[20]—delivers a seemingly never-ending urban expanse, only barely slowed by the limit edge where land meets water. Yellow vehicles cluster at the bottom of the image, near the bus station. Two major roads intersect at this point in a V, further creating the sense of movement outwards. Burtynsky mounted his camera to a drone in order to take in as much of the cityscape as possible. This hovering, bird's-eye view further accentuates the feeling of the city's expanse. It is a precise balance of far and near, of a scene we recognize—that is, a densely populated city—from a vantage that is likely unfamiliar, even for those who know Lagos well.

The Carrara mural (pages 149–151) presents us with a detail view of a mountain's quarried marble face, rather than an oblique overview. There is no main ground and no horizon, no way in or out, only a deep ledge of stone, as though we are in the mountain. The orange excavator is the only guide to position and relative size. This all produces conflicting impressions: the mountain continues to be extensively, methodically mined, and yet it remains majestic, ancient, enduring. At 3.7 × 7.3 metres, the mural brings the size of this natural resource to the fore, and in particular the scale of the quarrying operation.

The brightly coloured vertical coral wall in Indonesia (pages 173–175) presents a view that not even scuba divers can see: to the naked eye it appears as a monochrome blue-grey, due to the site's eighteen-metre depth. Burtynsky worked with twelve divers and specialized high-intensity underwater flash units to expose each section, a painstaking endeavour that took several days. The resulting image reveals a rich and biodiverse microcosm, a world independent of humanity and as yet unaffected by warming sea temperatures. The area is, however, very much at risk, as there is already extensive bleaching of coral reefs off the east coast of Australia.[21]

A swath of old-growth forest in Cathedral Grove, British Columbia (pages 167–169), provides a more familiar sight. Amid a lush, verdant scene of fir, cedar, spruce, and hemlock trees, and ferns, lichen, and moss, Burtynsky positions us as viewers with our feet firmly on the ground. We are small relative to the towering trees, and the sky is barely visible through the canopy. It appears untouched, prehistoric, and yet forests like these have been heavily logged since the nineteenth century.

To balance all the absorbing visual detail of the murals, which is almost too much to take in, the artists have created complementary experiences of these places, to animate and extend the sense of context. These film extensions, as Baichwal and de Pencier have termed them, accessed via an app, present a range of related vignettes, ways to further bring the particularities of a certain place to life.

In the case of the Carrara mural, one of the three film extensions expands the sense of topographical and industrial scale, while the other two highlight human activity. In the first, a drone view pulls away from the point at which the mural images were made, gradually revealing the full extent of the quarry and its system of roads, the artists three small specks in orange safety vests and yellow hard hats. The mountain's massive size and the craggy topography of the surrounding region are perhaps the most surprising, reinforcing the place as majestic and also deeply altered by humanity. The other two film extensions bring people to the fore—quarry workers and sculptors, actors in this extraction system. They reveal how the massive marble sections are cut, and how the raw material is then transformed into mass-market wares in a nearby sculpture studio (replicas of Michelangelo's *David*, at any size!); they make the labour required, both to extract the stone and to transform it for the market, very real. In as much as these systems seem remote to us, operating at a scale that can be difficult to fathom, the film extensions are a twofold reminder of human involvement: that these systems originate with and are perpetuated by humans, but also that the impact is a human one, and that, from wages to changing habitats, there are human stakes.

These murals with film extensions convey how the artists are actively grappling with what they are seeing—with the cumulative effect of our activity, and how the scale of it stretches credulity. It is also in these works that the artists' collaboration lives most clearly: a marriage of scale and detail, still and moving, abstraction and narrative.

The Immediate View

The newest territory de Pencier, Baichwal, and Burtynsky have ventured into as part of *The Anthropocene Project* is the realm of experiential technology. Virtual reality and augmented reality have gained much traction in recent years. The *New York Times,* for instance, has begun to explore both VR and AR as new modes of visual storytelling, citing an immersive, visceral quality as well as a "provocative explanatory value."[22]

Though uses of VR and AR have proliferated recently—in worlds as disparate as gaming, real estate, Holocaust education, and art historical research—they are still technologies whose main uses are very much in development.[23] At the moment, VR and AR seem to be valued for their ability to convey proximity and provoke a bodily, instinctual response. One of the goals, paradoxically, is to produce the sense of an unmediated experience, a greater directness. Where aerial views derive much of their power from the unusual remove they offer, AR and VR derive their power from an equally unusual sense of immediacy.

Debuting in these exhibitions are three AR installations, which present experiences of confiscated ivory tusks, a northern white rhinoceros, and a Douglas fir tree at or near actual scale. All highlight species threatened by extinction due to intensive human activity, another key marker of the Anthropocene; in these cases, poaching and logging are the primary causes. Both *AR #2, President Kenyatta's Tusk Pile, April 28, Nairobi, Kenya* (2016, pages 184–185) and *AR #4, Sudan, The Last Male Northern White Rhinoceros, Nanyuki, Kenya* (2016, pages 178–179) aim to bring viewers close to that which no longer exists: in the latter case a single animal, standing in for an entire species; in the former case a set of objects, standing in for the living animals to whose bodies they were once connected. In both cases the effect is deeply poignant. The scale of the tusk pile is striking, at roughly six metres in diameter and five metres in height. The activity needed to recover, confiscate, and catalogue the tusks also becomes evident, as viewers can examine markings on each one. The AR of the northern white rhinoceros, Sudan, is the only installation that also includes movement and audio: viewers can see him move and hear the quiet sound of his breathing. The animal's death in March of 2018 further underscores the ephemeral nature of the experience.

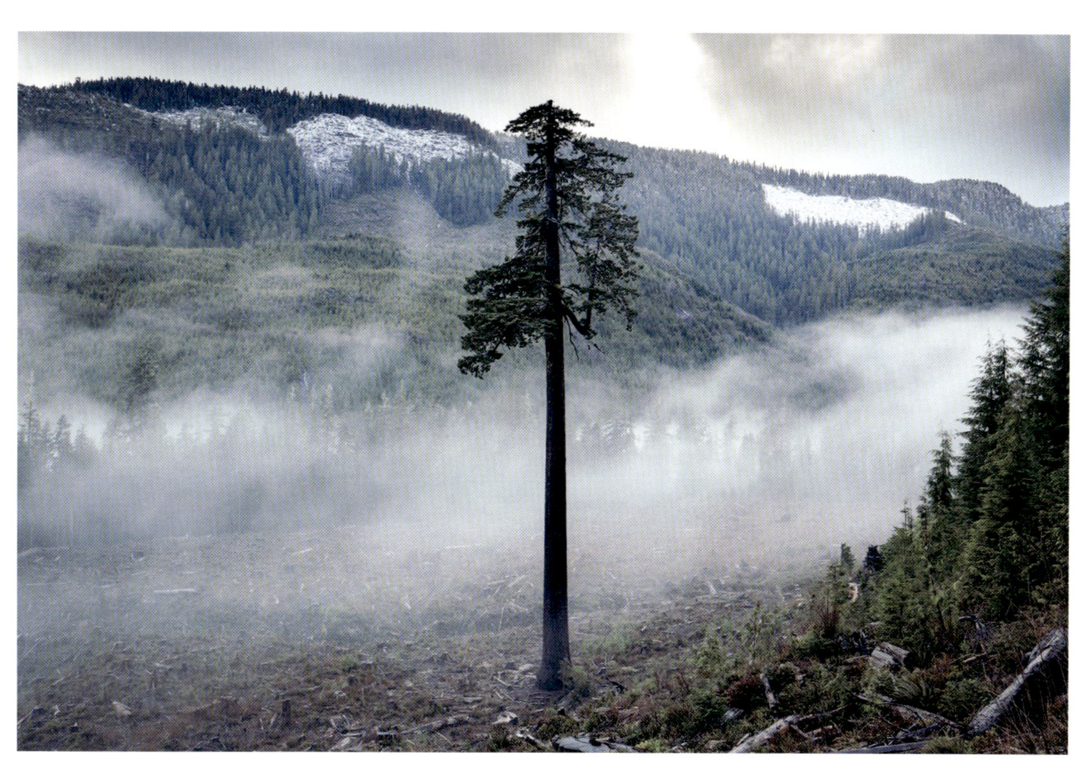

TJ Watt, *Big Lonely Doug,* February 21, 2016,
Ancient Forest Alliance.

As a new and complex type of image that conveys the most convincing three-dimensionality to date—built from thousands of individual images photogrammetrically mapped onto a virtual volume (a process, incidentally, that has evolved from the method of using aerial views to measure the distance between points on the ground, among other things)—AR is a compelling choice for this project. These installations push the long-standing use of photographs or films to document and memorialize people, animals, or places whose conditions are undergoing profound, irrevocable change. Though very different in form from a photograph or a film, these installations are nonetheless documents in the classical sense: their power resides in the very fact of what they present.

The third AR installation, *AR #3 Big Lonely Doug, Vancouver Island, British Columbia, Canada* (2016, pages 98–99), represents the second-tallest Douglas fir tree in Canada, one that stands sixty-six metres high, alone in an otherwise clear-cut forest outside Port Renfrew, an area that has seen an intense series of protests against the forest industry over the past few decades. Big Lonely Doug, as this single tree has come to be known, is thought to be about one thousand years old. According to Harley Rustad, this tree, saved by logger Dennis Cronin in 2011, has become "a symbol that is doing more to raise awareness about the cutting of old growth on Vancouver Island than any protest, march, or barricade."[24]

It is heady to consider that we can now "transport" Big Lonely Doug anywhere in the world and deliver the experience of standing at the foot of its massive trunk, twelve metres in circumference. What will it mean to view this tree in downtown Toronto, surrounded by high rises (and standing in an art gallery building renovated in 2008 with much Douglas fir), or in Ottawa, across the river from the Parliament Buildings—the first two venues for the *Anthropocene* exhibition? How will having Big Lonely Doug suddenly, silently present in our cities change the view? As Baichwal has stated: "A shift in consciousness can be the beginning of change."[25] In the experience of these artworks, there remains the possibility of greater awareness, a chance for us to deploy human ingenuity not only for destructive, extractive, narrowly self-interested purposes, but to shift what the landscape can mean to us. As philosopher Tristan Garcia has argued, the primary impulse to try to locate ourselves from above is essentially one of hope, an existential impulse to attempt to understand ourselves and, ultimately, to take responsibility.[26] In a distinct, yet related way, by attempting to erase distance, the AR installations stem from the same impulse.

For the artists, a single medium alone is no longer capable of conveying the necessary urgency: both collaboration and multiple media are essential. *The Anthropocene Project*, in all its facets, is the latest chapter in the ongoing attempt to galvanize broad support not only for better stewardship of the earth's resources but also for a new relationship to the planet through aesthetic experience. The astronauts of Apollo 8 were the first to see earth from space, the first to see the physical boundary of its atmosphere. The photographs they made in 1968, and those made since, have brought into view the strange vulnerability of a planet that many of its inhabitants have experienced for millennia as encompassing, omnipotent, limitless. Through the work of Burtynsky, Baichwal, and de Pencier, we are similarly confronted with a world we inhabit but cannot easily see. The experience of the places we shape with our actions and appetites, but may never lay eyes on, attest that, as a species, our power is, if not unlimited, certainly enough to prove decisive. The artists want us to carefully consider how to wield that power in the present.

Sophie Hackett is the Curator of Photography at the Art Gallery of Ontario and the co-curator of the *Anthropocene* exhibition.

1. Jennifer Baichwal, "Making of 'Watermark,' An Immersive Look at Water," *The Creators Project*, April 22, 2014, https://creators.vice.com/en_us/article/kbnwy9/making-of-watermark-an-immersive-look-at-water-video.
2. Darrell Varga, "On True Meaning(s) and the Impossibility of Documentary in the Films of Jennifer Baichwal," *Brno Studies in English* 39, no. 2 (2013): 56.
3. Baichwal.
4. The gigapixel essays and the 360° virtual reality films are not part of the exhibitions, but part of an online platform, *Anthropocene Interactive.*
5. De Pencier, in this volume, 206.
6. See Nicholas Mirzoeff, "Visualizing the Anthropocene," *Public Culture* 26, no. 2 (2014), and Heather Davis and Etienne Turpin, "Art & Death: Lives Between the Fifth Assessment and the Sixth Extinction," in Heather Davis and Etienne Turpin, eds., *Art in the Anthropocene* (London: Open Humanities Press, 2015), 3–29.
7. Denis Cosgrove and William L. Fox link this to a neurological capability for abstraction and imagination in *Photography and Flight* (London: Reaktion Books, 2010), 9–12. Tristan Garcia links this to the social sciences and defines it as a primary impulse of self-knowledge in "Le point de vue décollé," in Angela Lampe, ed., *Vues d'en Haut* (Metz: Centre Pompidou-Metz, 2013), 393–405.
8. Beaumont Newhall, *Photography 1839–1937* (New York: Museum of Modern Art, 1937), plate 84.
9. Gaëlle Morel, "Margaret Bourke-White," in Angela Lampe, ed., *Vues d'en Haut* (Metz: Centre Pompidou-Metz, 2013), 262–269.
10. See, for example, William L. Thomas, ed., *Man's Role in Changing the Face of the Earth* (Chicago: University of Chicago Press, 1956).
11. Ansel Adams and Nancy Newhall, *This Is the American Earth* (San Francisco: Sierra Club, 1992), 36.
12. Lewis Mumford, "Introduction," in E. A. Gutkind, *Our World from the Air* (Garden City, New York: Doubleday & Co. and the British Institute of Sociology, 1952), vi. Mumford also worked with the National Film Board of Canada on a six-part documentary film series, *Lewis Mumford on the City,* in 1963.
13. Cosgrove and Fox, 86. See also Stewart Brand, "Photography Changes Our Relationship to Our Planet," Smithsonian Photography Initiative, https://web.archive.org/web/20080530221651/http://click.si.edu/Story.aspx?story=31.
14. Carol Payne, *The Official Picture: The National Film Board of Canada's Still Photography Division and the Image of Canada, 1941–1971* (Montreal: McGill-Queen's University Press, 2013), 3.
15. It is interesting to note that Lorraine Monk, who selected the photographs for *Canada: A Year of the Land,* is credited on the title page as executive producer, a role more typically associated with filmmaking than book editing, and likens her approach to sequencing as akin to film editing.
16. Eight of Hunter's photographs were included in *Canada: A Year of the Land.*
17. "George Hunter: Photographer of the Industrial Scene," Art Gallery of Ontario, April 30, 2010, http://ago.ca/exhibitions/songs-future-canadian-industrial-photographs-1858-today.
18. Joel K. Bourne Jr., "California's Pipe Dream," *National Geographic,* April 2010, https://www.nationalgeographic.com/magazine/2010/04/plumbing-california/.
19. Edward Burtynsky, interview with the author, March 23, 2018.

20. Edward Burtynsky, quoted in Raffi Khatchadourian, "The Long View," *The New Yorker,* December 19–26, 2016, https://www.newyorker.com/magazine/2016/12/19/edward-burtynskys-epic-landscapes.

21. See "Reef Health," Great Barrier Reef Marine Park Authority, Australia Government, June 29, 2017, http://www.gbrmpa.gov.au/about-the-reef/reef-health.

22. See Marcelle Hopkins, "Pioneering Virtual Reality and New Video Technologies in Journalism," *New York Times,* October 18, 2017, https://www.nytimes.com/2017/10/18/technology/personaltech/virtual-reality-video.html, and Graham Roberts, "Augmented Reality: How We'll Bring the News into Your Home," *New York Times,* February 1, 2018, https://www.nytimes.com/interactive/2018/02/01/sports/olympics/nyt-ar-augmented-reality-ul.html.

23. One particularly compelling example is the *New Dimensions in Testimony* project, where survivors of the Holocaust are filmed and recorded to produce interactive hologram-like images. Designers of these experiences indicate that they didn't want viewers "to think there was technology around them at all." See Thomas McMullan, "The Virtual Holocaust Survivor: How History Gained New Dimensions," *Guardian,* June 18, 2016, https://www.theguardian.com/technology/2016/jun/18/holocaust-survivor-hologram-pinchas-gutter-new-dimensions-history.

24. Harley Rustad, "Big Lonely Doug," *The Walrus,* September 19, 2016, https://thewalrus.ca/big-lonely-doug/.

25. Baichwal, in this volume, 202.

26. Tristan Garcia, "Le point de vue décollé," in Angela Lampe, ed., *Vues d'en Haut* (Metz: Centre Pompidou-Metz, 2013), 404–405.

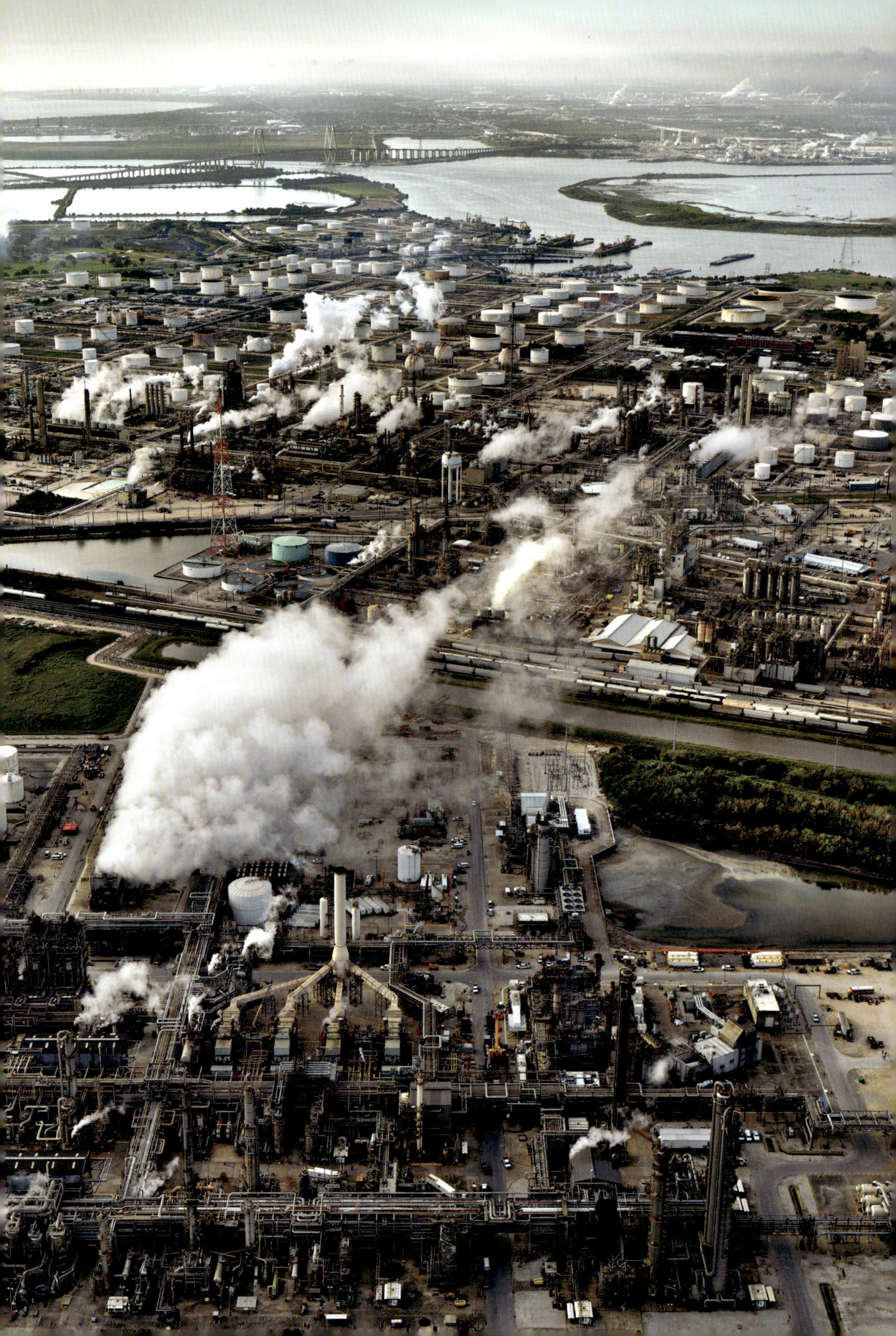

The Anthropocene and Its "Golden Spike"

Colin Waters & Jan Zalasiewicz

A little over eleven thousand years ago, the earth came out of the last of many glacial phases of the Ice Ages, and warmed into the latest of many warm interglacial phases. It was the expression of this phase of climate transition that geologists chose to distinguish by giving it the name of the Holocene epoch, as part of the Geological Time Scale. Over the succeeding eleven millennia, global climate was broadly stable, certainly by comparison with the roller-coaster trajectory of glacial-phase climate, as was sea level, once polar ice had finished melting some seven thousand years ago. In this time of relative stability of both the earth's geography and of the chemical cycles of carbon, nitrogen, and phosphorus that underpin life, settled human civilization developed and flourished.

When the Industrial Revolution started in the mid-eighteenth century, the situation that underpins this stability began to change. The burning of coal to supply the energy needs of growing European and North American economies, and a global billion-strong population, began to cause atmospheric CO_2 levels to rise towards the end of the nineteenth century. In the early to mid-twentieth century came a transition to the use of oil and natural gas, further perturbing the carbon cycle, and innovations such as the use of motor vehicles and the Haber-Bosch process began to flood the earth's surface with reactive nitrogen in the form of nitrogen oxides (NO_x) and ammonia-based fertilizers respectively, while phosphorus was mined from rock and spread over the land surface.

By the mid-twentieth century, when the earth's population had reached 3 billion, there began the postwar "Great Acceleration" of population, energy use, industrialization, technological development, and globalization.[1] It coincided with nearly two decades, starting in 1945, during which radioactive elements were distributed around the planet as a consequence of escalating atmospheric detonations of nuclear devices. By the end of the century, atmospheric CO_2 levels were a third above pre-industrial levels and climbing steeply. Surface reactive nitrogen and phosphorus levels had doubled. Landscape was being reshaped, not least with half a trillion tons of concrete, and new chemical products such as persistent pesticides and plastics were becoming widely disseminated on land and in the sea. The biosphere, too, was being remoulded by increasing extinctions, myriad species invasions, and the increasing domination of domesticated over wild species. Global temperatures and sea level rates of increase began to rise above their long-term baselines. Our planet is now on a new trajectory: from peak interglacial conditions into something yet hotter, together with an array of ongoing physical, chemical, and biological changes, many of which have no precedent in the earth's 4.6 billion year history.

OPPOSITE PAGE
Edward Burtynsky, *Petrochemical Plants, Baytown, Texas, USA* (detail), 2017.

The term "Anthropocene" was coined in 2000 and subsequently published by Paul J. Crutzen,[2] an atmospheric chemist and Nobel laureate, as a means of communicating this change from the relatively stable world of the Holocene to one in which humanity has altered the earth state into a new unstable one. The term found wide popularity, and developed diverse meanings across the sciences, humanities, and arts.

Its meaning as a part of Earth history, i.e., as a *geological* Anthropocene, had been intimated by Crutzen and his colleagues, but without consideration of the complex technical and formal aspects that typically surround newly proposed units of the Geological Time Scale. A geological investigation was initiated in response to the growing use of *Anthropocene* in scientific literature. It involved the inception of the Anthropocene Working Group (AWG) of the Subcommission on Quaternary Stratigraphy (SQS) in 2009, itself just one part of the main decision-making body, the International Commission on Stratigraphy (which is overseen in turn, and any decisions ratified—or not—by the International Union of Geological Sciences). The AWG's task has been to determine the merit of the Anthropocene as a potential new unit of the Geological Time Scale (more formally known as the International Chronostratigraphic Chart), by which earth's history is subdivided into a hierarchy of eons, divided in turn into eras, periods, epochs, and ages.

At the International Geological Congress in Cape Town in August 2016, the AWG announced its preliminary findings.[3] The Anthropocene, they found, was stratigraphically real. It should be defined at epoch level (such that the Holocene epoch has terminated, as Paul J. Crutzen had suggested). It should commence about the mid-twentieth century, broadly coincident with the Great Acceleration. This level was chosen not through historical significance (in which, arguably, the Industrial Revolution might be considered the key factor), but because the Anthropocene *as geology* is represented by strata, and these can be more effectively defined and demarcated as a globally synchronous physical unit by an array of stratal markers appearing in the mid-twentieth century. These markers include such things as artificial radionuclides, as well as the plastics and persistent organic pollutants mentioned earlier. The geological effects of the Industrial Revolution, by contrast, developed slowly and/or time-transgressively across the globe. Work is now underway to develop a proposal that the Anthropocene be formalized within the Geological Time Scale, using protocols that have been applied to its other time units (such as the Pleistocene, Jurassic, and so on).

For a unit of the Geological Time Scale to be established and ratified, the established protocols are complex and lengthy. They involve selection of a "primary stratigraphic marker," that is, a signal within strata such as the appearance of a fossil species or a geochemical change. Other "secondary markers" should be found that could help in the global recognition (termed by geologists "stratigraphic correlation") of the boundary where the primary marker is absent. The sharpness of these signals is important when choosing a reference boundary within strata in a specific part of the world—a "Global Boundary Stratotype Section and Point" (GSSP), more colloquially known as a metaphorical "Golden Spike"—which would demarcate the base of the Anthropocene. A single section is unlikely to represent all parts of the world, and so selection of a number of *auxiliary* stratotypes is helpful, in which the same level is represented by other geological signals elsewhere, with enough signals in common to allow cross-referencing between them. Putting all this together is a substantial task.

A range of potential stratigraphic markers is available to define the base of the Anthropocene.[4] Many are anthropogenic in origin, entirely novel in Earth history, and hence have not been used previously to define boundaries of ancient geological time units. A marked upturn in plutonium and radiocarbon radioisotope abundance, approximately in 1952 and 1954 respectively, is broadly coincident with a downturn in stable carbon and nitrogen isotope ratios that reflect the extraordinary growth in hydrocarbon use and nitrogen fertilizer manufacture, and these signals may be recognised across most sedimentary environments. Hence, this fallout radioactive "bomb spike" from thermonuclear atmospheric tests may prove suitable as a primary marker.[5] There are other airborne signals, such as fly ash, nitrates, and to a lesser extent sulphur and sulphates; carbon dioxide and methane concentrations (preserved directly within the annual snow layers that build polar ice); persistent organic pollutants such as the pesticides dichlorodiphenyltrichloroethane (DDT), aldrin, and dieldrin; and heavy metals seen as both increases in lead abundance in sediments and ice, and as changed lead isotope ratios (that record the global use of tetraethyl lead additives in gasoline). These signals provide widespread, additional means of global correlation of the Anthropocene boundary in strata as secondary markers.

Microplastics, too, have seen almost global distribution throughout lakes, river systems, and across the oceans, and these may be a suitable marker, supplementing the several billion tons of plastic waste accumulating in anthropogenic deposits such as landfill sites.[6] Biological signals, typically used as the primary marker for the majority of GSSPs in ancient rock successions, show an intrinsic complexity typical of young successions (a complexity that is often "smoothed out" in the million-year-scale time scales recorded by ancient strata). There has been a marked upturn in extinction events over the last five hundred years (potentially leading towards the earth's sixth mass extinction event over the past half billion years), and extensive transformation of entire ecosystems, such as increasing rates of coral bleaching events over the last forty years as sea surface temperatures rose, but there is no obvious single globally distributed species disappearing in the mid-twentieth century that might serve as a marker for the Anthropocene. However, there have been many regional influxes of invasive species (both intentional and accidental) and introductions of domesticated species; individual examples are typically time-transgressive across the planet, but where their distribution in time and space is well enough understood, they may become useful time markers to help locate the Holocene/Anthropocene boundary.

Within the more "proximal" Anthropocene strata upon which humans live and have more direct and larger influences, the geological succession is commensurately more complex in time and space. Think of the structure of a growing megacity as an enormous and intricate "urban stratum," for example, made both of the above-ground structures and of below-ground meshworks of tunnels, pipes, and cables, penetrating both rock and natural sediment and the masses of anthropogenic waste material; this waste is widely spread as layers of "artificial ground," and also locally concentrated as the growing urban landfill sites, that may be several tens of metres thick and occupy several square kilometres. This kind of novel geology is too complex and irregular to serve well as host for a "Golden Spike" (where regularity and temporal continuity is demanded), yet it is of enormous importance in characterizing and understanding the Anthropocene.

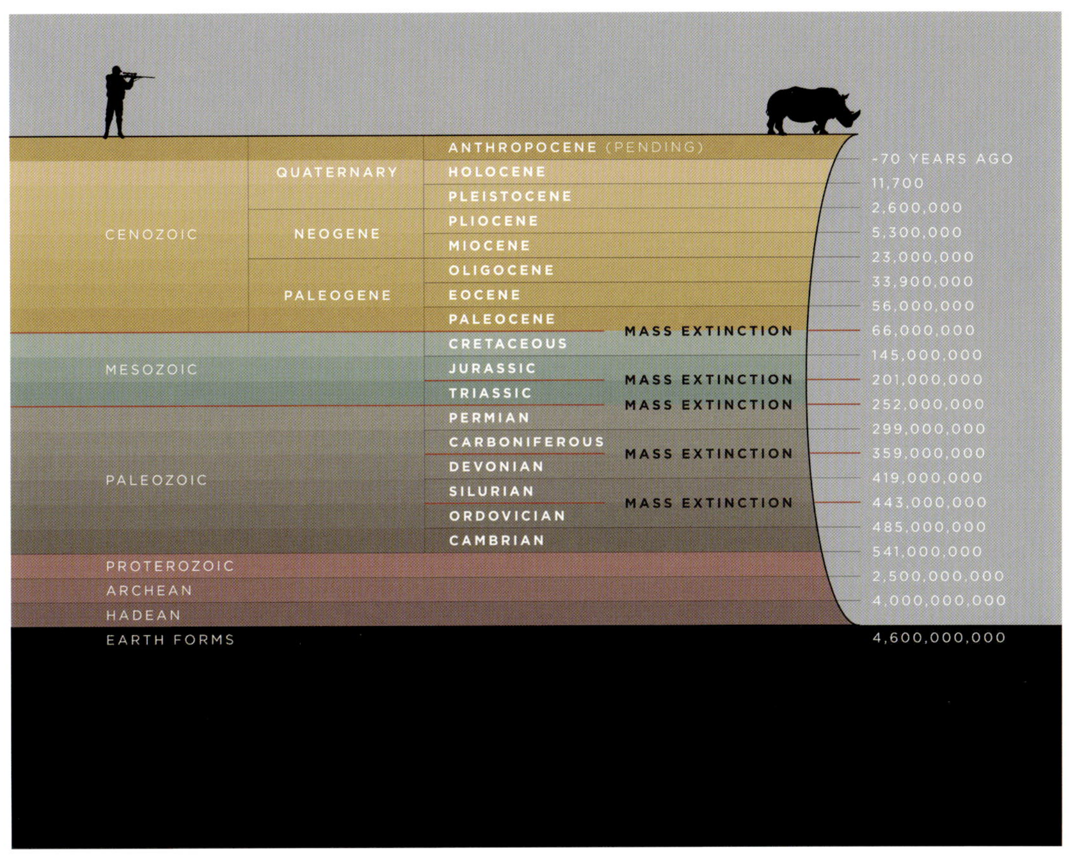

The geological evolution of planet earth since its formation 4.6 billion years ago. Infographic by Barr Gilmore.

One way to think of this terrain is through the technosphere concept,[7] being a new "system" of the earth, related to and emerging from the biosphere, lithosphere, hydrosphere, and atmosphere, with its own emergent properties and dynamics. Humans and their social structures are part of the technosphere—perhaps more caught up within its momentum than constructing and directing it—as are all of our technological constructs and the waste products spilling out of these. One aspect of this new planetary system is the sheer mass of its physical components. A recent approximate assessment[8] put this at some 30 trillion tons; averaged, that is about fifty kilograms of human-processed matter (about one kilo of which is concrete) per square metre of the globe. About a third of this mass is made up of the world's urban areas. Much of this refashioning has taken place in the Anthropocene, driven by an enormous increase in energy output (by some estimates, humans have expended as much or more energy since the mid-twentieth century as they did in all of the earlier Holocene). Processes within the technosphere are commonly complexly interlinked. Increasing rates of deforestation, turning the land over to agriculture, and reshaping the landscape through excavating mineral resources or construction of transport networks have contributed to enhanced rates of erosion of soils and shallow sediments. Construction of some 58,500 large dams across the major river systems of the planet, at a rate of about two a day during the Anthropocene, has impounded some 3 trillion tons of sediment produced through the enhanced erosion within resultant reservoirs. This starves river deltas, many of which are the location of megacities, of much of their accumulation of deposits, facilitating greater subsidence and enhancing the risk of flooding, already elevated through rising sea levels.

The rapid growth in bulk of the technosphere is one characteristic of the Anthropocene. Another is rapid and accelerating diversification in the form of its myriad physical components—artifacts of various kinds that one might name "technofossils" because many have high fossilization potential, being durable and resistant to decay. Technofossils might be regarded as a form of trace fossil, akin to an animal burrow or footprint, though hugely more diverse (most animals produce only a few forms of traces, while humans have designed and constructed many millions of different types of artifact). Of significance to the stratigraphy of the Anthropocene is the evolution of these technofossils, many orders of magnitude faster than biological evolution and so providing time markers that can help characterize Anthropocene strata. Some whole classes of material are essentially confined to the Anthropocene, such as most plastics, and within these, some technofossils represent yet finer time divisions, such as PET (polyethylene terephthalate) bottles appearing in 1973 and compact discs in 1982; an iconic evolutionary lineage is that of the mobile phone, with several generations evolving and becoming obsolescent over a couple of decades.

This accelerating growth in the bulk and diversity of the physical technosphere is at the expense of the biosphere, not least as burgeoning domesticated animals (e.g., the modern broiler chicken, now unable to survive without constant human maintenance) are as much a part of the rapidly evolving technosphere as they are of the biosphere. The displacements and extinctions of organisms within the "natural biosphere" are producing their own complex signal in contemporary strata.

The technosphere and its products represent one means of wider analysis of the Anthropocene, and this forms part of the context for constructing the proposal to formalize this term within the Geological Time Scale. The protocol for this is clear. From among such a wide range of environments, a small number of potentially representative sites require the acquisition of more systematic and comprehensive datasets, with correlation established between sections, to allow selection of a candidate GSSP and several auxiliary stratotypes. A Holocene–Anthropocene boundary should be widely recognizable in practice, being globally distributed across a wide range of environments, and the initial assessment undertaken will help find optimal stratotype locations among these environments.

Why go to such lengths to define the Anthropocene? Defining it formally is a means of giving precise meaning, within the geological framework of "deep" earth time, to this rapidly evolving assemblage of processes and their material products. This kind of precision of meaning in itself helps scientific communication, both within and beyond the earth sciences. Also, the geological signals of the Anthropocene in very young strata allow comparison, like for like, with the signals of environmental change preserved in ancient strata, allowing human impact to be measured on a planetary scale of time and space, and not just as a phenomenon of human history or ecology. This allows the scale, rate, and novelty of Anthropocene change to be compared with change in prehistoric times. Some of the comparisons are striking: for instance, the rate of rise of atmospheric carbon dioxide in the Anthropocene is over a hundred times faster than its rate of rise as the last glaciation event gave way to the Holocene—a rate of rise that is itself considered rapid on a geological scale. Other changes are modest geologically. The amount of sea level rise so far in the past century—some twenty centimetres—is geologically trivial, and even the metre-scale rise projected for the next century and already detected as acceleration of rate of rise[9] is exceedingly modest. But, given the recent human propensity for building megacities on coastlines, even this modest change will have large consequences for our species, and so has fostered considerable interest in the Anthropocene concerning the implications for the International Law of the Sea.[10] The term has also become a synonym for extensive global environment change, much broader in concept than climate change, which can be used to express and contextualize issues such as apprehension about the human health implications of these changes.[11]

The geological Anthropocene hence gives a particular and, in some respects, an objective context within which to understand current global change. Its perspective encompasses and integrates not just climate change (which, still being in its early stages, remains only a modest part of the characterization of the Anthropocene) but also a more diverse array of global physical, chemical, and biological changes. Interestingly, the geological definition complements almost perfectly the Earth System science definition, as originally developed by Paul J. Crutzen and his colleagues,[12] even though the forms of analysis differ greatly. This helps give confidence that the Anthropocene has considerable and substantial reality as a planetary phenomenon.

The quest for a formal geological definition of the Anthropocene is neither a simple task nor one that is certain to be accepted by the governing bodies, given the considerable institutional barriers to *any* change to the Geological Time Scale, let alone one that involves as many novel and difficult issues as the Anthropocene. Nevertheless, the rigorous research required to formulate a proposal on the Anthropocene, at the very least, is helping to illuminate the reality of striking change to the geohistorical trajectory of our planet.

Colin Waters is an honorary professor in the Department of Geology at the University of Leicester, and secretary of the Anthropocene Working Group, with a central role in coordinating activities of the Working Group members. He recently retired as a Principal Mapping Geologist at the British Geological Survey, where he specialized in geological mapping of the U.K. and parts of the Sahara Desert, and stratigraphical analysis, principally of the Carboniferous and Anthropocene.

Jan Zalasiewicz is Professor of Palaeobiology at the School of Geography, Geology, and the Environment at the University of Leicester, and chairs the Anthropocene Working Group. Formerly a field geologist and paleontologist at the British Geological Survey, he now teaches a range of geological subjects to undergraduate and postgraduate students, and researches ancient deep-ocean strata of the early Paleozoic and environmental change in the Quaternary Ice Ages, including now the more contemporary geology of the Anthropocene.

1. Will Steffen, Wendy Broadgate, Lisa Deutsch, Owen Gaffney, and Cornelia Ludwig, "The Trajectory of the Anthropocene: The Great Acceleration," *The Anthropocene Review* 2, no. 1 (2015): 81–98.

2. Paul J. Crutzen, "Geology of Mankind," *Nature* 415, no. 23 (January 2002): 23.

3. Jan Zalasiewicz, Colin N. Waters, Colin P. Summerhayes, Alexander P. Wolfe, Anthony D. Barnosky, Alejandro Cearreta, Paul J. Crutzen, Erle Ellis, Ian J. Fairchild, Agnieszka Gałuszka, Peter Haff, Irka Hajdas, Martin J. Head, Juliana A. Ivar do Sul, Catherine Jeandel, Reinhold Leinfelder, John R. McNeill, Cath Neal, Eric Odada, Naomi Oreskes, Will Steffen, James Syvitski, Davor Vidas, Michael Wagreich, and Mark Williams, "The Working Group on the Anthropocene: Summary of Evidence and Interim Recommendations," *Anthropocene* 19 (September 2017): 55–60.

4. C. N. Waters, J. Zalasiewicz, C. Summerhayes, et al., "The Anthropocene Is Functionally and Stratigraphically Distinct from the Holocene," *Science* 351, no. 6269 (January 2016): 137.

5. C. N. Waters, J. P. M. Syvitski, A. Gałuszka, et al., "Can Nuclear Weapons Fallout Mark the Beginning of the Anthropocene Epoch?" *Bulletin of the Atomic Scientists* 71, no. 3 (2015): 46–57.

6. J. Zalasiewicz, C. N. Waters, J. Ivar do Sul, et al., "The Geological Cycle of Plastics and Their Use as a Stratigraphic Indicator of the Anthropocene," *Anthropocene* 13 (2016): 4–17.

7. P. K. Haff, "Technology as a Geological Phenomenon: Implications for Human Well-being," in *A Stratigraphical Basis for the Anthropocene*, eds. C. N. Waters, J. Zalasiewicz, M. Williams, M. A. Ellis, and A. Snelling (London: Geological Society, Special Publications): 301–309.

8. J. Zalasiewicz, M. Williams, C. N. Waters, et al., "Scale and Diversity of the Physical Technosphere: A Geological Perspective," *The Anthropocene Review* 4, no. 1 (2016): 9–22.

9. R. S. Nerem, B. D. Beckley, J. T. Fasullo, et al., "Climate-change-driven accelerated sea-level rise detected in the altimeter era," *Proceedings of the National Academy of Sciences* (February 2018).

10. Davor Vidas, "The Anthropocene and the International Law of the Sea," *Philosophical Transactions of the Royal Society A* 369, no. 1938 (2011): 909–925.

11. S. Whitmee, A. Haines, C. Beyrer, et al., "Safeguarding Human Health in the Anthropocene Epoch: Report of The Rockefeller Foundation–Lancet Commission on Planetary Health," *Lancet* 386, no. 10007 (2015): 1973–2028.

12. W. Steffen, R. Leinfelder, J. Zalasiewicz, et al., "Stratigraphic and Earth System Approaches in Defining the Anthropocene," *Earth's Future* 4, no. 8 (2016): 324–345.

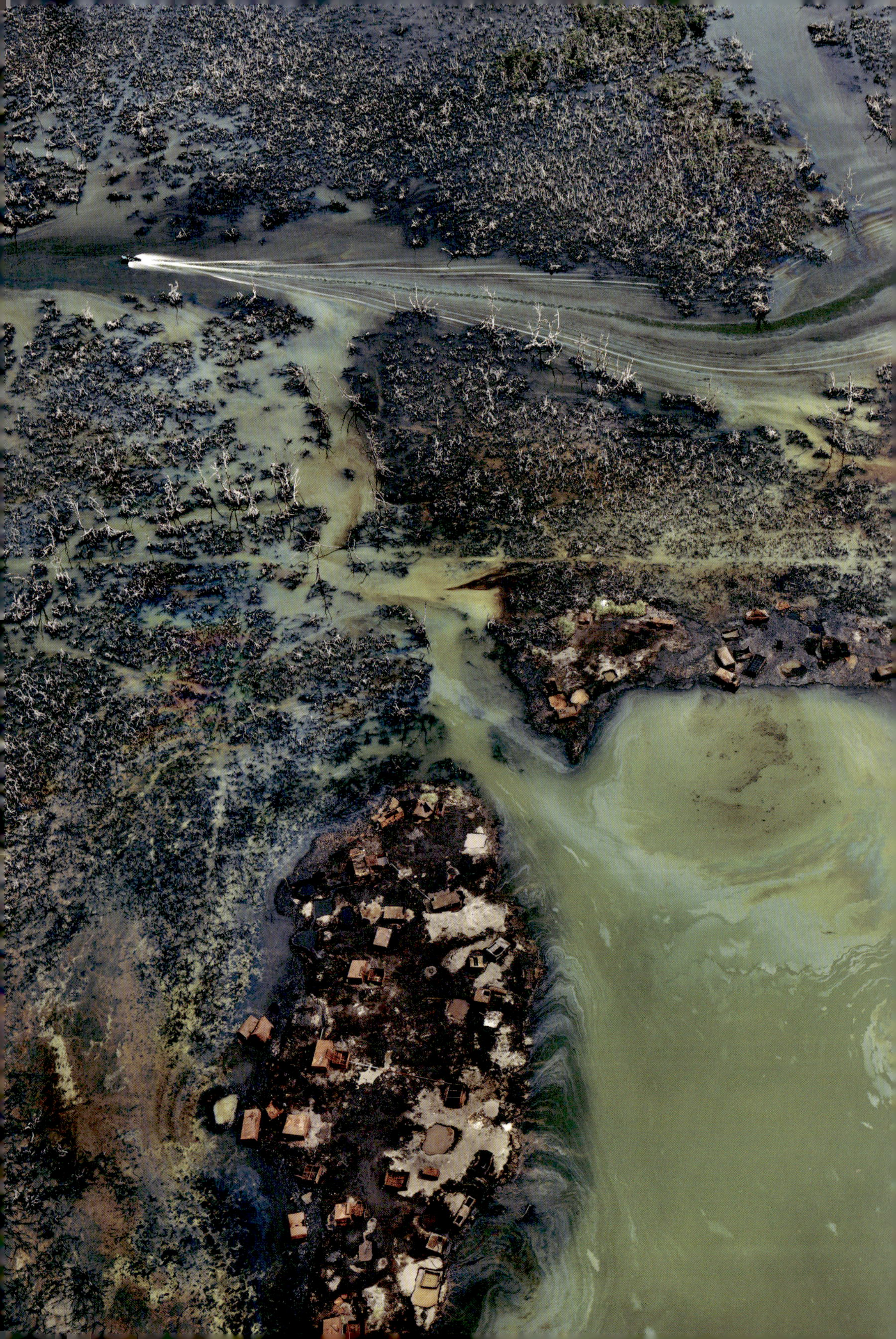

"How Anthropo-scenic!": Concerns and Debates about the Age of the Human

Karla McManus

There is a lot of talk about the Anthropocene. As we circle the drain, our greatest thinkers and provocateurs—including scientists and artists—are debating, disagreeing, and disrupting the concept of the new geological epoch, the literal Age of Human. When did it start? people ask. Who is responsible? Can we prevent more change or at least slow it down? Will naming it help? Some of this is just semantics, of course, a debate about the details, not the concept itself. Yet from another perspective, semantics—that, at times, dismissive term used to describe the practice of nitpicking—is at the very heart of the cultural understanding of human communication and meaning-formation. Words have power, after all, and those who wield them even more. When we argue about the basic meaning of a term—for example, does the *anthropo* imply human responsibility is equally shared?—we are making cultural decisions about what the Anthropocene will mean to people for years, maybe even centuries, to come. This is an important dialogue, even if it can sometimes come across as petty or schismatic. Perhaps the biggest question of all is one rooted in the dynamics of power: Who gets to decide the answers to these questions?

These debates are guided by the question of what it means to be human on the planet, when we have irrevocably altered the balance of the earth's systems: the hydrosphere, the biosphere, the pedosphere, the atmosphere. These changes are beyond physical; they have affected our cultures and histories, driven wars and alliances, shaped our economies, and influenced the way people have been enslaved, subjugated, and exploited, from the age of agriculture to the age of steam and the microchip. We have disrupted and altered our global climate, acidified our waters, triggered the mass extinction of biodiversity, transformed ecosystems that we rely on for life, polluted vast areas of land and water, all the while reproducing at a rate that requires more and more resources that are becoming scarcer.[1] The complexities and interdependencies of these systems are only now becoming understood, even while we change them irrevocably. How do we even begin to draw a line or plant a spike to say, "This is when it all started"?

OPPOSITE PAGE
Edward Burtynsky, *Oil Bunkering #4, Niger Delta, Nigeria* (detail), 2016.

Scientists, that vast group of specialists that includes everyone from astrophysicists to molecular biologists and those who study all the bits of space and time found in between, are having their own debates. According to historical practice and convention, a new geological age must be measured at the chronostratigraphic level, in the layered rock and fossil sediments of the earth. That is how we have always defined an epoch, be it the Holocene, which began some 12,000–11,000 years ago (give or take) at the end of the last Ice Age, or the earlier Pleistocene, which began over 2.5 million years BP (before present). Yet even the boundary of those two epochs, located both within the Quaternary Period, is debated by some scientists, just as the "Golden Spike," the physical reference point in the strata from one particular strategic location used to demarcate a particular shift in geological time (officially known as a Global Boundary Stratotype Section and Point), is disagreed upon by many.[2]

Is the evidence strong enough to declare a whole new geological epoch rather than, say, an age, a smaller unit of measurement? The atmospheric chemist Paul J. Crutzen certainly thinks so. Since 2000, when he famously made the declaration that human changes to the global environment had led us into a new geological epoch, the Anthropocene, Crutzen has written numerous essays in collaboration with biologist Eugene F. Stoermer, chemist Will Steffen, and historian John R. McNeill, to articulate his definition of the Anthropocene. While not the first to use this term, Crutzen has based the definition of the phenomenon on his own speciality of atmospheric science, informed by the clearly measurable increase of carbon dioxide particles in the earth's atmosphere since 1800.[3] As early as 2002, Crutzen wrote that "because of the anthropogenic emissions of carbon dioxide, global climate may depart significantly from natural behaviour for many millennia to come."[4] And, like a good scientist, he roots the solution (or at least mitigation) of the problem in "improved technology and environmental management, wise use of Earth's remaining resources, control of human and of domestic animal population, and overall careful treatment and restoration of the environment—in short, responsible stewardship of the Earth System."[5]

The Anthropocene Working Group (AWG) has agreed that the idea has merit. Made up of a group of scientists and historians, and led by geologist Jan Zalasiewicz, the AWG was formed in response to the free adoption of the term, which had begun to seep into the literature without any real scientific analysis or evidence of its boundaries or scope. This led to the founding of the AWG in 2009, an international research group tasked with considering the evidence for the new epoch for their parent body, the International Commission on Stratigraphy. In 2016, at the 35th Annual International Geological Congress in Cape Town, South Africa, the AWG announced that the concept of a new geological epoch was a sound one in a majority vote on the preliminary findings. Yet they have only agreed to date the epochal change to around 1950, rooting it in a measurable "bomb spike" of radionuclide signals from the detonation of the first atomic weapons in 1945, and the testing that followed.[6] As of today, the AWG continues to finalize their findings, but the fundamental debate over whether there is sufficient evidence to support the epochal change has been resolved, for now. For the scientists involved, the concept of the Anthropocene and the research on its existence is solid enough to move forward with the proposal to the International Commission on Stratigraphy.

Among the AWG's findings are some interesting reflections on the fundamental questions I posed in my introduction. The members of the AWG are pragmatically aware that the concept of the Anthropocene and its adoption are loaded with more than purely scientific impact. For example, a 2015 article by geographers Simon Lewis and Mark Maslin suggested that the AWG consider two possible dates for the beginning of the Anthropocene: 1610 and 1964. The latter date proposed corresponds with the peak of radionuclide levels in the atmosphere, such as plutonium isotopes and caesium-137 (an isotope that is not naturally occurring), and it is close to the AWG's final recommended start date. The earlier date corresponds with a significant and measurable dip in carbon levels in the earth's atmosphere that was brought on by the mass population decline from the genocide and ecocide that followed European arrival in the Americas. Further evidence for this date, according to Lewis and Maslin, including the discovery of New World plants in Old World sediments, would meet the AWG's requirements for stratigraphic evidence.[7] Such a date would also, as cultural theorists Heather Davis and Zoe Todd have recently written, "name the problem of colonialism as responsible for contemporary environmental crisis."[8] In response to Maslin and Lewis, the Anthropocene Working Group argues in favour of the most conservative and measurable date possible, laying out in an article the flaws with their two proposed dates (in short, not enough global physical evidence and placed not at the beginning of the upswing). The AWG justifies this decision by acknowledging that:

> We are aware of the narratives that may be built around the Anthropocene, and how these may be influenced by boundary choice. However, we suggest that the positioning of a stratigraphic boundary should simply be pragmatically and dispassionately chosen, by the same manner in which all earlier stratigraphic boundaries were chosen, to allow the most effective practical division between what would then become (by definition) Anthropocene and pre-Anthropocene strata and history. Such a choice would, we consider, *be the best guarantee that wider discussion is solidly founded on the best factual basis available* [emphasis added].[9]

Yet, as much as the AWG may want to control the "narratives" of the term, every day more and more journal articles and monographs, newspaper and web stories, are published asking us to consider the answers to these questions from outside the hard sciences, with their focus on the stratigraphy in the fossils and rocks that make up the surface of the earth. I point this out not to call into question the science or even the logic of this decision, but only to acknowledge, as the AWG does here, that scientific decisions are enmeshed in cultural and social decisions. For the AWG, choosing the most widely accepted date by majority consensus of its members—a date that reflects planetary synchronous change and is therefore easily defended by scientific evidence—is more important than choosing a date that reflects the truly catastrophic beginnings of human-altered environmental change, such as the global spike in CO_2 levels first held as responsible by Paul J. Crutzen or the impact of mass genocide in the Americas.[10]

Such cultural roots of the Anthropocene suggest that the AWG's mid-century marker—what Crutzen, McNeill, and Steffen refer to as the beginning of the "Great Acceleration,"[11] which includes an increase in everything from population growth and fertilizer usage to energy consumption and water usage—is only the culmination of many centuries of human-nature conflict. There is no question that the bombing of Hiroshima and Nagasaki, and the years of nuclear testing that followed, have most definitely altered life on earth, detectable in the stratosphere as well as in the flesh, most famously in the breast milk of mothers of both human and bovine extraction.[12] This bodily and stratigraphic evidence of human-driven geophysical change may be as good as any marker for illustrating how human ambition in the form of technology, militarization, and globalization have transformed life on earth and the potential future of the planet.

For some scientists, the ideal usage of the Anthropocene concept is as "a rhetorical device designed to shape thinking in the realms of government, business, and civil society,"[13] one that would allow for the premise of a human-driven solution based in geo-engineering or a "business-as-usual" mentality that privileges scientific determinism over societal change.[14] As a result of this knee-jerk positivism, many social scientists have been grappling with the ontological meaning and epistemic uses of "the charismatic mega-category"[15] that is the Anthropocene. To further understand how the ways in which we define and use the term shape its meaning, geographer Jamie Lorimer has identified five categories of deployment, from the most pragmatic to the most creative. His questions, with my interjections, go like this: the scientific question (is it fact?), intellectual zeitgeist (that makes sense to me!), ideological provocation (Capitalocene, Eurocene, Manthropocene, etc.), new ontologies for thinking through the human-natural dialectic (decolonization, post-humanism, and new materialism are such examples), and, finally, the idea of science fiction or speculative fiction as an entree into all of the many implications and expressions of possibility, good and bad, in the Human Epoch.[16]

Many of us from the social sciences, humanities, arts, and political spheres feel that we should also have a voice in this discussion, a discussion that for the non-scientific community is not only rooted in questions of evidence but is also to be found, perhaps even more so, in human experience and history. We understand this term in what we as individuals perceive, remember, and experience emotionally, bodily, visually in the environments we know. The widespread understanding that we are living in and through a transformative period in the history of the earth, brought on by human activity at a scale never thought possible, has created a sense of urgency that is at odds with the very nature of geological thinking, which sees history occurring over billions of years and evidence as set in stone.

Indeed, the challenge of perceiving the Anthropocene as *both* geological and historical points to one of the main criticisms levelled at the concept, which places the emphasis on *anthropo,* as, at its most simplistic, this nomenclature seems to suggest that humans aren't in fact part of the natural world but, exceptionally, outside it.[17] Such a conflict, funnily enough, has been noted by many as the reasons we find ourselves in the Anthropocene epoch. The hubris of naming an epoch after ourselves has not been missed by some, who point out that, even if we started it, the effects of humanity on the geological scale will far outlive us as a species.[18] Driven by the desire to enrich human life with the wealth of the earth, we have over-extracted, over-consumed, over-produced with no thought about the planet as an ecosystem or the climate as alterable by our actions. For how can you really make any impact on what's written in stone? For Dipesh Chakrabarty, this is a conflict rooted in the different perceptions of Anthropocene time, what he describes as based on either human-centred or planet-centred thinking. Chakrabarty puts it clearly that, "the Anthropocene requires us to think on the two vastly different scales of time that Earth history and world history respectively involve: the tens of millions of years that a geological epoch usually encompasses versus the five hundred years at most that can be said to constitute the history of capitalism."[19] For Chakrabarty, such planetary crisis leads us to the limits of historical understanding.

There is some irony in this—throughout much of our recent history, human actions on the earth were perceived as insignificant. How could we make any dent in the wealth of nature, when its riches were so overabundant and bountiful? Yet since the nineteenth century, geographers, historians, and anthropologists have been sounding alarms about how human activity in one part of the world was impacting another. This awareness of planetary environmental impact caused by human activity was a minority one, certainly, yet the evidence was there. As historian Richard Grove has pointed out, many of these early observers of human-driven ecological system change were working in the context of Empire and witnessing the drastic environmental changes brought about by colonialist human activity on the globe, from over-trapping to plantation agricultural practices to deforestation and soil desiccation.[20] The imperialist history of the Anthropocene, and the rise of what historian Jason W. Moore has called "cheap nature," is one of the most disturbing realities of the time and place we live in. Inextricably linked to the early modern period of globalization, when the Atlantic slave trade helped to fuel a new and complex international commodities system called capitalism, the Anthropocene concept co-evolves with human attitudes to capital-N Nature as a resource to be extracted, sometimes even reducing humans to the status of natural product to be exploited.[21] This is why some cultural theorists, Moore and Donna Haraway among them, have argued against the Human Age, instead proposing the Capitalocene, a term meant to acknowledge how nature has been turned into resource by the powerful economic forces of capitalism, rooted in human social history and extraction ideology.[22]

This is not to say that before capitalism, humans lived in an Edenic state with nature. As historian William Cronon so succinctly put it in his 1995 essay, "The Trouble with Wilderness: Or, Getting Back to the Wrong Nature"—an article that sparked a surprising backlash against what was perceived as a relativist and postmodern understanding of wilderness—the modern concept of nature as free from civilization is described as a cultural myth shaped by human desires and fantasies. As he put it:

> If we allow ourselves to believe that nature, to be true, must also be wild, then our very presence in nature represents its fall. The place where we are is the place where nature is not. [...] To the extent that we celebrate wilderness as the measure with which we judge civilization, we reproduce the dualism that sets humanity and nature at opposite poles. We thereby leave ourselves little hope of discovering what an ethical, sustainable, honourable human place in nature might actually look like.[23]

This antithesis, or even antipathy, between nature and humanity, which places humans at some apex and positions nature as the lowly *other,* is also found in the discourse of the Anthropocene. If nature is something that we humans are in charge of, as both spoilers and stewards, then it stands to reason that we are in charge of coming up with the big solutions. Haraway argues that the concept of the Anthropocene threatens to be much too big and to end badly, relying as it does on an imaginary dead end: Progress.[24] For if we think that we are separate from nature, and therefore the masters of the earth's systems, then we can surely solve any of its problems by applying a little ingenuity and elbow grease. This very Enlightenment-based approach to the Anthropocene, as Alan Mikhail has pointed out, centres the human figure as both responsible for the cause and the solution to our geophysical crisis.[25] By doing so, we reject the interconnectedness of human, animal, soil, water, and air, and simplify all the complexity that we know is at work in the earth's system. We also risk ignoring the historical complexities that come into play between Enlightenment ways of thinking and the pre-modern. Mikhail puts it this way: "The problems with the Enlightenment Anthropocene narrative are the problems of the Enlightenment narrative—too clean a break between the modern age and everything before it, a divide between nature and human, a discourse of homogenizing universality too often used in the service of exerting differential power over the weak."[26]

These pre-Enlightenment ideas were never completely forgotten in our current times. Popularized in the 1970s, the Gaia hypothesis, developed by James Lovelock with Lynn Margulis, proposed that the earth functions as a complex system seeking homeostasis through the balancing of its spheres: pedo-, bio-, hydro-, and atmos-. This theory brought many non-scientists to an understanding of how human actions on soil or water might have larger reactions beyond the local pond or forest.[27] But before such useful scientific conceptual models reached the popular imagination, we have had other models of thinking in homeostasis: the religious philosophies of animism, which posited a soul in all things,[28] the ecological thinking of Romantics such as Henry Thoreau,[29] the land ethic of mid-century conservationists like Aldo Leopold, the multi-species poetic activism of Rachel Carson,[30] and the radical bio-centric philosophy of deep ecologists who seem to suggest humans don't deserve to exist if they can't live in harmony with the planet.[31] Many of these models reflect a troubled disassociation between gender, race, capital, and environment, from Thoreau's misanthropy to deep ecology's refusal to acknowledge its romantic primitivism.[32] Nevertheless, from a purely generous and generative perspective, such ways of thinking have helped to open up discussions of how the environment can be thought of outside an instrumentalist relationship to humans. Recently, the call to respect and reflect on traditional ecological knowledges of Indigenous peoples from the land that is now Canada, and from around the world, has offered a valuable way to think about land stewardship and the complexities of environmental rights, justice, and human/non-human animal relationships. Aims to decolonize and/or indigenize the Anthropocene are informed by what Kyle Powys Whyte calls the post-Apocalyptic moment in which Indigenous people find themselves today, caught in an ever-changing environment brought on by the settler-colonial state, the "dystopian future."[33]

For sociologist of science Bruno Latour, the emphasis on human-driven solutions ignores that "the point of living in the epoch of the Anthropocene is that all agents share the same shape-changing destiny, a destiny that cannot be followed, documented, told, and represented by using any of the older traits associated with subjectivity or objectivity."[34] Anthropologist Anna Tsing has called our existence an "interspecies relationship," which would not exist without microbes, spores, and multi-species landscapes that we live in.[35] So how can we break away from our particular and peculiar speciesist thinking and instead reflect and react in symbiosis with the earth (and ourselves) to find solutions that lead to fundamental and epistemic changes in our relationship to the Anthropocene?

Some have called for a greater engagement with the imagination of the Anthropocene to think through more radical and hopeful ways toward the future.[36] Science fiction writers have been especially good at helping to imagine, to *picture* even, the Anthropocene, even before the term was used.[37] Think of Aldous Huxley's dystopian bio-technological future in *Brave New World* (1932), or John Brunner's polluted and corporate America in *The Sheep Look Up* (1972), or Octavia E. Butler's *Xenogenesis* series (1987, 1988, and 1989), which imagines a post-nuclear earth remade habitable by an alien race. These stories bring up possibilities of the future, helping us to imagine the disturbing follow-throughs of problems we have created, including climate change, nuclear risk, over-population, and genetic engineering; sometimes they even propose radical solutions. Other times the dystopian qualities of these stories can seem too far-fetched or easily disassociated from the reality of today (human-alien hybrids?). The best of science fiction in and about the Anthropocene may be those examples that help us to envision a kind of multi-species environmental justice that eco-critical scholar Ursula Heise sees as central to understanding the biodiversity loss of the sixth mass extinction without descending into despair.[38]

As a temporal structure that seems to suggest futurity over the present, the Anthropocene makes a difficult subject of visual representation.[39] Visual culture theorist Nicholas Mirzoeff believes that the Anthropocene visuality embodies a conundrum, as it enforces the authority of a Western aesthetic mode rooted in imperialism—both past and present—and the conquest of nature rather than activating resistance. For Mirzoeff, Anthropocene visuality produces a kind of *anesthetic*, which numbs our awareness of the actual physical conditions of the Anthropocene, and must be fought against through the reclaiming of the imagination, using a countervisuality, that "claims the right to see what there is to be seen and name it as such: a planetary destabilization of the conditions supportive of life, requiring a decolonization of the biosphere itself in order to create a new sustainable and democratic way of life."[40]

Setting aside the kind of Anthropocene visualizations that we are so used to—from graphs of the accelerating levels of CO_2, to satellite images of the earth showing time-lapses of melting glaciers, to the ubiquitous polar bear on an ice floe—we need to creatively imagine new ways that art can engage the Anthropocene.[41] For Heather Davis and Etienne Turpin, writing in the introduction to their edited collection *Art in the Anthropocene*, artists working with the ideas of the Anthropocene can help to erase the thinking that seems to impose a teleological chronology of forward marching progress. Art, they write, helps us to understand "the various times that pass through ours, the ways in which time can bend and elongate, and how time is written into our bodies, composing the relations we have to all the other things around us."[42]

How do we represent the *now* of the Anthropocene? Only by acknowledging that the present is always partly of the past and simultaneously of the future can we appreciate the way that Anthropocene time has changed our relationship to the planet. Artists such as Edward Burtynsky, Jennifer Baichwal, Nicholas de Pencier, and many others, have approached these challenges from a variety of perspectives, using education, activism, aesthetics, and cutting-edge technology to present their vision of the Anthropocene today. Using different scales and modes of visualizing the planetary history of human activity, *The Anthropocene Project* gives viewers a sense of what is at stake and invites us to reinvest in the future. *The Anthropocene Project* asks us to consider: What can images do to help us understand the current debates that are circulating about the brand-new epoch we have brought about?

While the concerns and provocations I have outlined above will continue to rage on, and new problems and possibilities will arise, this project offers, right now, a way to represent and experience the Anthropocene through the power of photography. Instead of dwelling on the impossibility of visualizing the vastness of climate change, or the limitations of photography as a form of populist realism,[43] let us instead put our minds and hearts to using images as a conduit into the possible. Photography offers the viewer an experience predicated on the conjunction of time and space and fact and emotion, shaped heavily by our imaginations. As curator Lucy Lippard eloquently writes, "The most powerful images present not just what they see; they are also haunted by what we don't see."[44] The future is haunting us all, as much as the past. Let us avoid what social theorist McKenzie Wark has decried as the relentless tendency to critique and privilege our little silos of Anthropocene knowledge and instead work towards "building forms of collaborative scientific, technical, intellectual, organizational, affective, and manual labour to confront the Anthropocene and find a path through the unstable time it announces."[45]

The Anthropocene is not going anywhere; we have all the time of our world.

Karla McManus is an art historian who specializes in the study of photography and the environmental imaginary. She is currently a Limited Term Assistant Professor at Ryerson University's School of Image Arts.

1. Anthony D. Barnosky et al., "Introducing the Scientific Consensus on Maintaining Humanity's Life Support Systems in the 21st Century: Information for Policy Makers," *The Anthropocene Review* 1, no. 1 (April 1, 2014): 78–109, https://doi.org/10.1177/2053019613516290.

2. Paul J. Crutzen and Will Steffen, "How Long Have We Been in the Anthropocene Era?" *Climatic Change* 61, no. 3 (2003): 251–257, https://doi.org/10.1023/B:CLIM.0000004708.74871.62; Will Steffen et al., "The Trajectory of the Anthropocene: The Great Acceleration," *The Anthropocene Review* 2, no. 1 (April 1, 2015): 81–98, https://doi.org/10.1177/2053019614564785; Colin N. Waters et al., "Can Nuclear Weapons Fallout Mark the Beginning of the Anthropocene Epoch?" *Bulletin of the Atomic Scientists* 71, no. 3 (January 2015): 46–57, https://doi.org/10.1177/0096340215581357; Andreas Malm and Alf Hornborg, "The Geology of Mankind? A Critique of the Anthropocene Narrative," *The Anthropocene Review* 1, no. 1 (April 1, 2014): 62–69, https://doi.org/10.1177/2053019613516291.

3. Will Steffen, J. R. McNeill, and Paul J. Crutzen, "The Anthropocene: Are Humans Now Overwhelming the Great Forces of Nature?" *Ambio* 36, no. 8 (2007): 614–621.

4. Paul J. Crutzen, "Geology of Mankind," *Nature* 415, no. 3 (January 2002): 23.

5. Crutzen and Steffen, "How Long Have We Been in the Anthropocene Era?" 256.

6. Jan Zalasiewicz et al., "The Working Group on the Anthropocene: Summary of Evidence and Interim Recommendations," *Anthropocene* 19 (September 1, 2017): 55, https://doi.org/10.1016/j.ancene.2017.09.001.

7. Simon L. Lewis and Mark A. Maslin, "Defining the Anthropocene," *Nature; London* 519, no. 7542 (March 12, 2015): 175.

8. Heather Davis and Zoe Todd, "On the Importance of a Date, or, Decolonizing the Anthropocene," *ACME: An International Journal for Critical Geographies* 16, no. 4 (December 20, 2017): 763.

9. Jan Zalasiewicz et al., "Colonization of the Americas, 'Little Ice Age' Climate, and Bomb-Produced Carbon: Their Role in Defining the Anthropocene," *The Anthropocene Review* 2, no. 2 (August 1, 2015): 123, https://doi.org/10.1177/2053019615587056.

10. A breakdown of the thirty-five members' votes was made public in a preliminary report published in 2017. While 28.3 voters agreed that 1950 should be the beginning of the Anthropocene, the majority (25.5) also agreed that a Global Stratotype Section and Point (the Golden Spike) was the best way to define the start of the Anthropocene. Interestingly, it was here that the group had the most debate: plutonium fallout received ten votes and radionucleides received four for a total of fourteen votes in favour of the "bomb spike." While this was by far the majority of votes for any particular "primary marker," it is revealing to note that six of the AWG's members abstained on this topic. Zalasiewicz et al., "The Working Group on the Anthropocene," 58–59.

11. Steffen et al., "The Trajectory of the Anthropocene."

12. Steven L. Simon, André Bouville, and Charles E. Land, "Fallout from Nuclear Weapons Tests and Cancer Risks: Exposures 50 Years Ago Still Have Health Implications Today That Will Continue into the Future," *American Scientist* 94, no. 1 (2006): 48–57.

13. Noel Castree, "The Anthropocene: A Primer for Geographers," *Geography; Sheffield* 100 (Summer 2015): 70.

14. Malm and Hornborg, "The Geology of Mankind?" 67.

15. Elizabeth Reddy, "What Does It Mean to Do Anthropology in the Anthropocene?" *Platypus: The CASTAC Blog,* April 8, 2014, http://blog.castac.org/2014/04/what-does-it-mean-to-do-anthropology-in-the-anthropocene/.

16. Jamie Lorimer, "The Anthropo-Scene: A Guide for the Perplexed," *Social Studies of Science* 47, no. 1 (2017): 117–142, https://doi.org/10.1177/0306312716671039.

17. Eileen C. Crist, "On the Poverty of Our Nomenclature," in *Anthropocene or Capitalocene? Nature, History, and the Crisis of Capitalism*, ed. Jason W. Moore (Oakland: PM Press, 2016), 16–17.

18. Dipesh Chakrabarty, "The Climate of History: Four Theses," *Critical Inquiry* 35, no. 2 (January 2009): 197–222, https://doi.org/10.1086/596640.

19. Dipesh Chakrabarty, "Anthropocene Time," *History and Theory* 57, no. 1 (March 9, 2018): 6, https://doi.org/10.1111/hith.12044.

20. Richard H. Grove, "Environmental History," in *New Perspectives on Historical Writing*, ed. Peter Burke, 2nd ed. (University Park: Pennsylvania State University Press, 2001), 268–269.

21. Jason W. Moore, ed., "The Rise of Cheap Nature," in *Anthropocene or Capitalocene? Nature, History, and the Crisis of Capitalism* (Oakland, CA: PM Press, 2016), 86–87.

22. Jason W. Moore, ed., *Anthropocene or Capitalocene? Nature, History, and the Crisis of Capitalism* (Oakland: PM Press, 2016).

23. William Cronon, "The Trouble with Wilderness: Or, Getting Back to the Wrong Nature," *Environmental History* 1, no. 1 (January 1996): 17.

24. Donna Haraway, "Staying with the Trouble: Anthropocene, Capitalocene, Chthulucene," in *Anthropocene or Capitalocene? Nature, History, and the Crisis of Capitalism*, ed. Jason W. Moore (Oakland: PM Press, 2016), 54.

25. Alan Mikhail, "Enlightenment Anthropocene," *Eighteenth-Century Studies* 49, no. 2 (January 29, 2016): 223, https://doi.org/10.1353/ecs.2016.0002.

26. Ibid, 226.

27. James E. Lovelock and Lynn Margulis, "Atmospheric Homeostasis by and for the Biosphere: The Gaia Hypothesis," *TUS Tellus* 26, nos. 1–2 (1974): 2–10.

28. David Pepper, *Modern Environmentalism: An Introduction* (New York: Routledge, 1996), 28.

29. Lawrence Buell, *The Environmental Imagination: Thoreau, Nature Writing, and the Formation of American Culture* (Cambridge, M.A.: Belknap Press of Harvard University Press, 1995).

30. Rachel Carson, *The Sea around Us*, rev. ed. (New York: Oxford University Press, 1961).

31. Arne Naess, *Ecology of Wisdom: Writings by Arne Naess*, 1st ed. (Berkeley: Counterpoint: distributed by Publishers Group West, 2010).

32. Ramachandra Guha, *Radical American Environmentalism and Wilderness Preservation: A Third World Critique* (Albuquerque: John Muir Institute for Environmental Studies and University of New Mexico, 1989); Peter C. Van Wyck, *Primitives in the Wilderness: Deep Ecology and the Missing Human Subject* (Albany: State University of New York Press, 1997); Juan Martínez Alier, *The Environmentalism of the Poor: A Study of Ecological Conflicts and Valuation* (New Delhi: Oxford University Press, 2005); Rob Nixon, *Slow Violence and the Environmentalism of the Poor* (Cambridge, M.A.: Harvard University Press, 2011).

33. Heather Davis, Etienne Turpin, and Zoe Todd, eds., "Indigenizing the Anthropocene," in *Art in the Anthropocene: Encounters Among Aesthetics, Politics, Environments and Epistemologies 2015* (London: Open Humanities Press, 2015), http://openhumanitiespress.org/Davis-Turpin_2015_Art-in-the-Anthropocene.pdf; Kyle Powys Whyte, "Our Ancestors' Dystopia Now: Indigenous Conservation and the Anthropocene," in *The Routledge Companion to the Environmental Humanities*, ed. Ursula K. Heise, Jon Christensen, and Michelle Niemann, 1st ed. (London: Routledge, 2017), 207; Kyle Powys Whyte, "Indigenous Climate Change Studies: Indigenizing Futures, Decolonizing the Anthropocene," *English Language Notes* 55, no. 1/2 (Fall 2017): 153–162.

34. Bruno Latour, "Agency at the Time of the Anthropocene," *New Literary History* 45, no. 1 (April 23, 2014): 15, https://doi.org/10.1353/nlh.2014.0003.

35. Anna Tsing, "Unruly Edges: Mushrooms as Companion Species," *Environmental Humanities* 1 (November 1, 2012): 144.

36. Kathryn Yusoff and Jennifer Gabrys, "Climate Change and the Imagination," *Wiley Interdisciplinary Reviews: Climate Change* 2, no. 4 (2011): 516–534. See also Greg Garrard, Gary Handwerk, and Sabine Wilke, eds., "Imagining Anew: Challenges of Representing the Anthropocene," special section of *Environmental Humanities Journal* 5 (2014).

37. Kate Marshall, "What Are the Novels of the Anthropocene? American Fiction in Geological Time," *American Literary History* 27, no. 3 (2015): 523–538, https://doi.org/10.1093/alh/ajv032; Andrew Milner et al., "Ice, Fire and Flood: Science Fiction and the Anthropocene," *Thesis Eleven* 131, no. 1 (December 2015): 12–27, https://doi.org/10.1177/0725513615592993.

38. Ursula K. Heise, *Imagining Extinction: The Cultural Meanings of Endangered Species* (Chicago: University of Chicago Press, 2016).

39. Peter Galison and Caroline A. Jones, "Unknown Quantities," *Artforum International* 49, no. 3 (November 2010): 49–51, 282.

40. Nicholas Mirzoeff, "Visualizing the Anthropocene," *Public Culture* 26, no. 2 (2014): 230, https://doi.org/10.1215/08992363-2392039.

41. For an interesting polemic calling on artists to work towards representing and engaging with the social and environmental catastrophe, see Rasheed Araeen, "Ecoaesthetics: A Manifesto for the Twenty-First Century," *Third Text* 23, no. 5 (September 2009): 679–684, https://doi.org/10.1080/09528820903189327; and the response by Mark A. Cheetham et al., "Ecological Art: What Do We Do Now?" *Nonsite.org*, March 21, 2013, http://nonsite.org/feature/ecological-art-what-do-we-do-now.

42. Heather Davis and Etienne Turpin, eds., "Art & Death: Lives Between the Fifth Assessment & the Sixth Extinction," in *Art in the Anthropocene: Encounters Among Aesthetics, Politics, Environments and Epistemologies 2015* (London: Open Humanities Press, 2015), 12, http://openhumanitiespress.org/Davis-Turpin_2015_Art-in-the-Anthropocene.pdf.

43. For further discussions of the limits and possibilities offered by the photographic medium, see essays by curator Bénédicte Ramade and art critic and historian T. J. Demos in Bénédicte Ramade, *The Edge of the Earth: Climate Change in Photography and Video* (London: Black Dog Publishing, 2016).

44. Lucy R. Lippard, *Undermining: A Wild Ride in Words and Images Through Land Use Politics in the Changing West* (New York: New Press, 2014), 171.

45. McKenzie Wark, "Social Theory for the Anthropocene," *The Futures We Want: Global Sociology and the Struggles for a Better World*, November 2, 2015, http://futureswewant.net/mckenzie-wark-anthropocene/.

WORKS

Itzurun Beach, part of Spain's Basque Coast UNESCO Global Geopark, is a unique landscape where geologic time is made visible through dramatic layers of sedimentary rock called "flysch." These turbidite fans, composed of fine, alternating deposits of shale and sandstone, are only possible along such coastal shorelines as the one pictured here, where sediment builds up on deep-sea floors adjacent to evolving mountain chains. When the sea floor is pushed up into land through tectonic forces, this jagged layering occurs (leaning from west to east; oldest to newest), a result of underwater avalanches that slide down the slopes of the seabed. As the materials rest on the ocean floor, different sedimentation speeds cause a gradation effect. Bigger particles, such as sand and other coarse material, settle first to form the ground layer that is then overlaid by increasingly finer particles,[1] resulting in the unique strata along this coast. The length of this eight-kilometre beach represents a timescale of approximately 50 million years, estimated to stretch from 100 million years ago to 50 million years ago. Of particular relevance to the stratigraphic consideration of the Anthropocene, Itzurun Beach contains a complete record of the boundaries between two geologic ages, the Cretaceous-Paleogene (K-Pg boundary) and the Paleocene-Eocene,[2] and indicators of their associated mass extinctions. While the K-Pg mass extinction is commonly believed to be the result of an asteroid impact (one most known for the extinction of non-avian dinosaurs), the Paleocene-Eocene Thermal Maximum (PETM) indicates a period of carbon-induced global temperature rise, which lasted approximately 200,000 years.[3] Evidence of this can be seen in the PETM section at Zumaia. The sequence indicates changes in the marine organisms of the time, including traces of deep-sea benthic foraminifera, which experienced a major extinction during this period of warming.[4] This ridged shoreline is also home to two noted Global Boundary Stratotype Sections and Points (GSSPs) within the Paleocene epoch, indicating the bases of the Selandian and Thanetian stages.[5] Commonly referred to as Golden Spikes, these stratotypes are the formal demarcation of a geological time unit. At time of writing, the Golden Spike of the Anthropocene has yet to be determined. Altogether, this stretch of flysch-covered coast in Basque country presents a unique microcosm of earth's history and potential trajectory; world-changing events are now visible only as thin layers of rock. Considering that agriculture was developed only some twelve thousand years ago,[6] the flysch at Zumaia serve as sobering reminders of the impact humanity has imposed on the planet in such a short time.

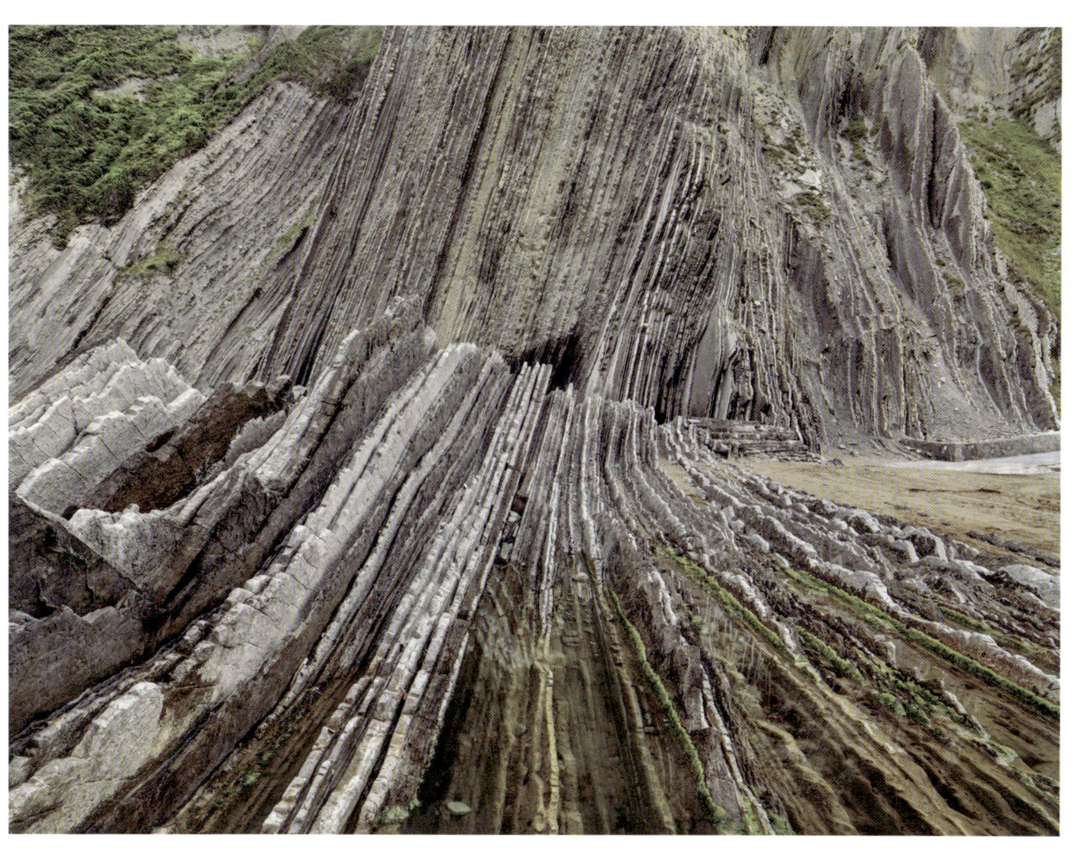

Edward Burtynsky, *Basque Coast #1, UNESCO Geopark, Zumaia, Spain*, 2015

The Dandora Landfill is among the largest of its kind in the world. Referred to as Nairobi's Municipal Dumping Site, the area receives industrial, agricultural, commercial, and medical waste, amounting to about 2,000 tonnes per day.[1] Due to its location next to the Nairobi River, any runoff is carried into the water system. It is estimated that nearly a million people live in the vicinity of the landfill.[2] Though steps were taken to decommission the landfill in 2012, it was never officially closed and continues to operate with no active replacement.[3] Residents work informally, sorting scrap by hand and selling it to newly built recycling plants on site. The mounds in these images, some 4.6 metres high, are composed primarily of less valuable plastic bags. In 1950, less than 2 million tonnes of plastics were manufactured globally per year.[4] By the early twenty-first century, this amount had reached 300 million tonnes per year.[5] The total cumulative amount of plastics produced by 2015 was calculated to be 5 billion tons, enough to cover the entire earth in plastic wrap. Too small for typical methods of filtration, microplastics are virtually ubiquitous in our environment, and are increasingly deposited in sediment layers, making them a key technofossil for the stratigraphic consideration of the Anthropocene.[6] In 2017, plastic bags were banned across Kenya, a move increasingly made by governments worldwide that want to reduce their nation's plastic footprints.

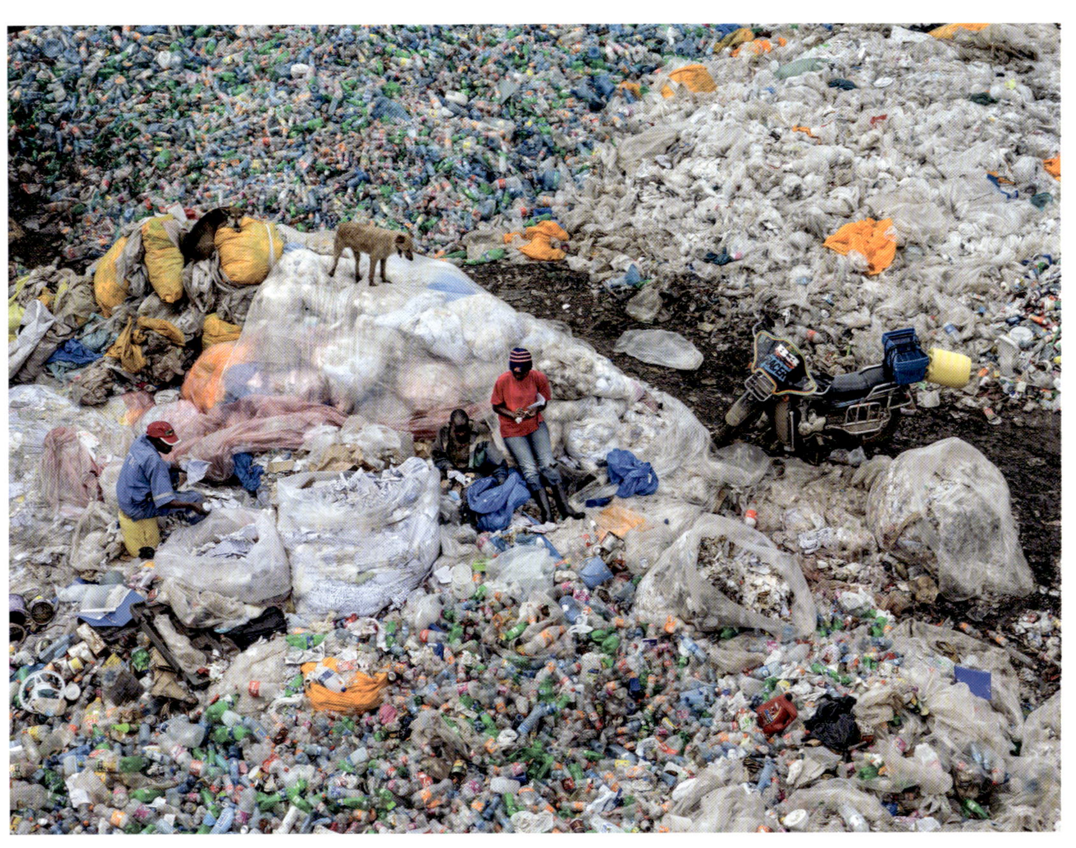

Edward Burtynsky, *Dandora Landfill #3, Plastics Recycling, Nairobi, Kenya,* 2016

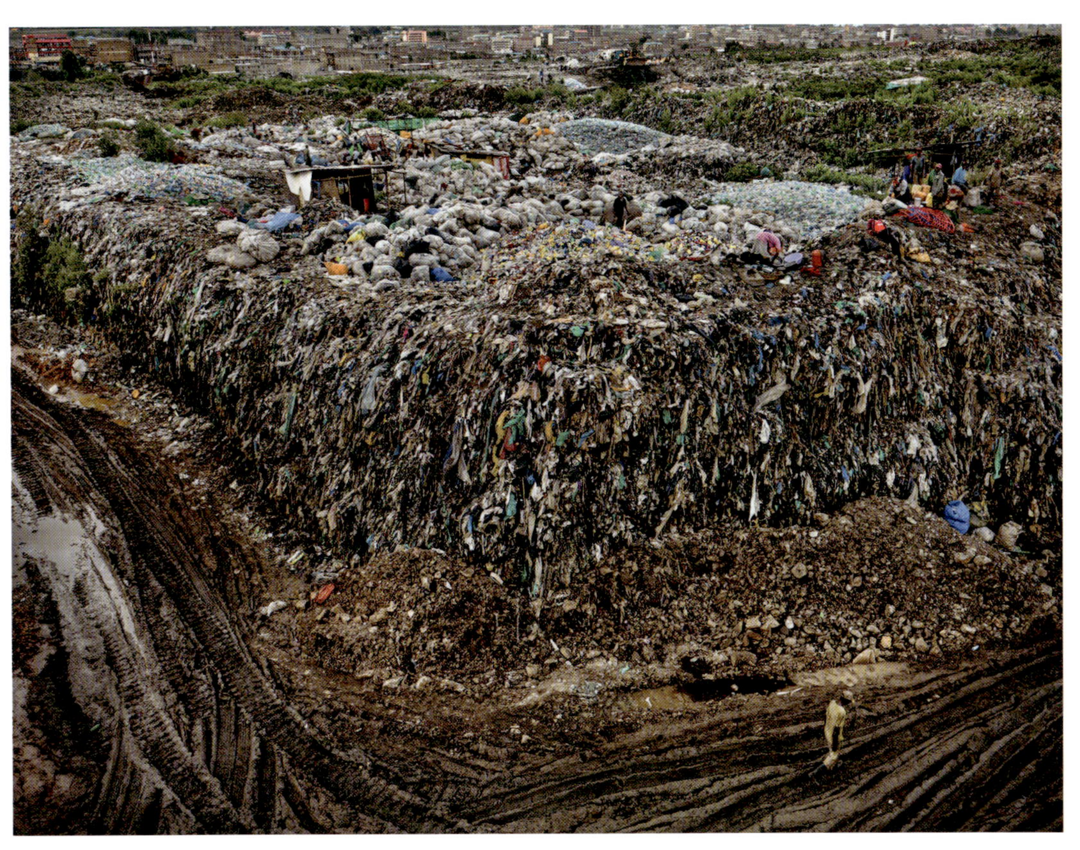

Edward Burtynsky, *Dandora Landfill #1, Nairobi, Kenya*, 2016

The Dandora Landfill was created in the 1970s and declared full more than a decade ago.[1] However, the municipal dumping site continues to function and provides the primary income source for many in its vicinity. An estimated six thousand people mine its fenceless grounds each day,[2] searching for metal, rubber, glass, plastics, and electronics for resale or purchase by recycling companies.[3] A 2007 study by the United Nations Environment Programme found fatally high levels of lead in soil samples bordering the landfill.[4] Nearly half of the 328 local children tested suffered from respiratory problems and exhibited lead concentrations in their blood that exceeded internationally accepted levels.[5]

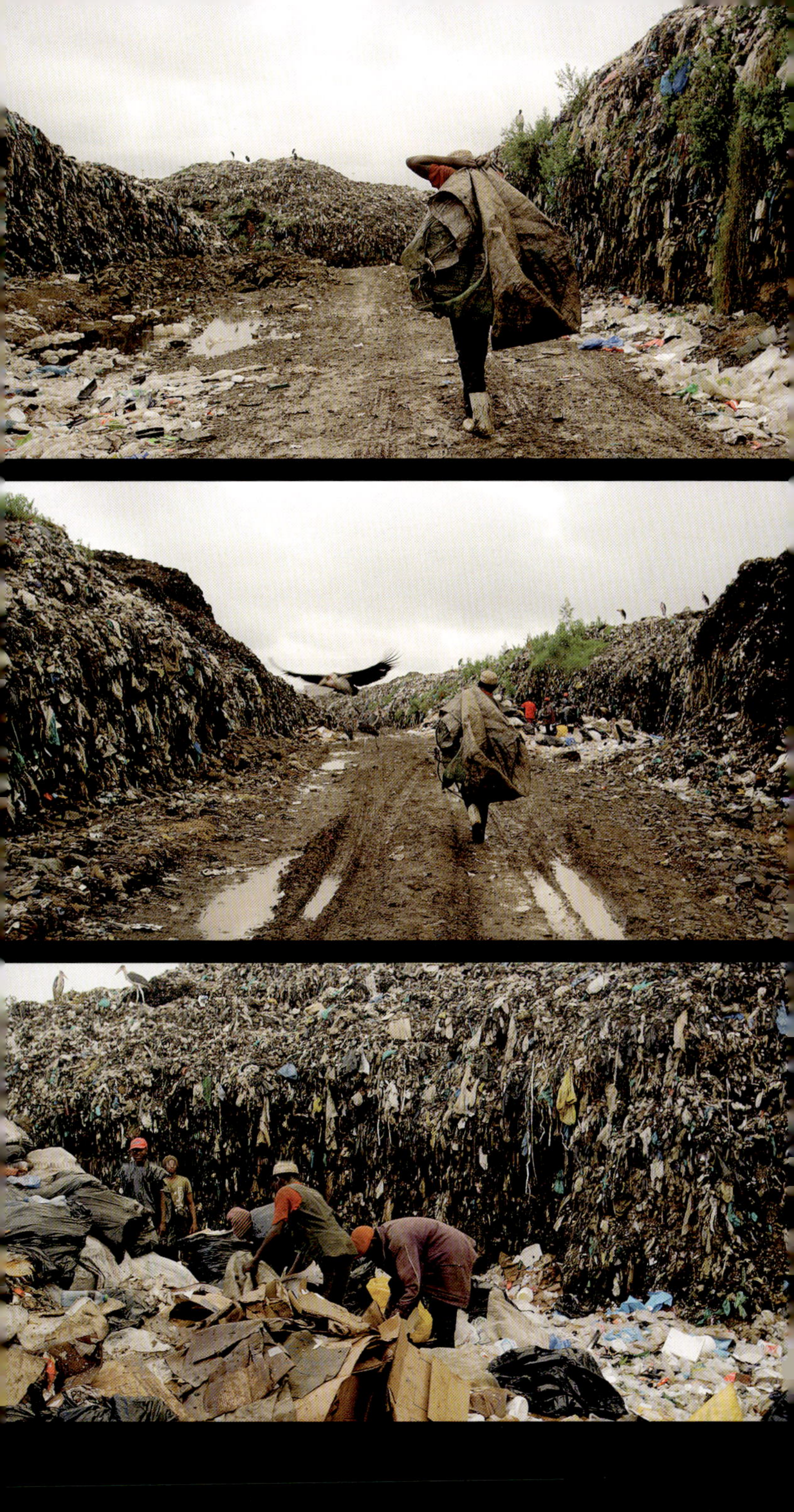

Jennifer Baichwal and Nicholas de Pencier, *Dandora Landfill, Nairobi, Kenya* (film stills), 2018

Technofossils are human-generated objects that, if preserved in the strata, will serve as future geological markers through which the Anthropocene epoch can be considered. Cement is one of the most significant technofossils that humanity will leave behind. Invented by the Romans, concrete became a fundamental, global building material during the mid-twentieth century. Since that time, enough concrete has been poured to coat the earth in a two-millimetre-thick layer of the material. From 1995 to 2015, rapid urbanization and population growth saw the production of more than half of the planet's total volume of concrete.[1] In sum, as a human-made material, concrete is unparalleled in quantity. The tetrapods pictured here make use of concrete to mitigate another human-generated problem: climate change. When used for shoreline protection, tetrapod seawalls allow water to flow around them. This disperses the energy of breaking waves that would otherwise crash against a flat wall, eroding the shoreline. Along China's coast, a region that comprises only 13 percent of China's total land area, but contributes 60 percent of the country's gross domestic product (GDP), continued erosion poses a serious threat.[2] Technological advancements in machinery during the early twentieth century led to increased sediment displacement, mainly flux, in most large rivers around the world.[3] By the 1950s, this sediment displacement began to reverse due to the rapid construction of dams, resulting in sediment load reduction below pristine conditions.[4] This reduction of sediment downstream frequently results in riverbank erosion, as well as a decline in the nutrients deposited in floodplain areas, thereby undermining the chemical base of the ecosystem.[5] Such sediment displacement is considered a global signal of the Anthropocene, as is delta subsidence, which began in the 1930s and is now a dominant warning signal for many coastal environments.[6] Simultaneously, seawalls are built to reclaim wetlands for urban and industrial expansion, resulting in a stark decline in biodiversity.[7] The need for more land to support the world's most populous country, combined with environmental threats to existing settlement, has resulted in the construction of new seawalls covering over 60 percent of the total length of China's coastline.[8]

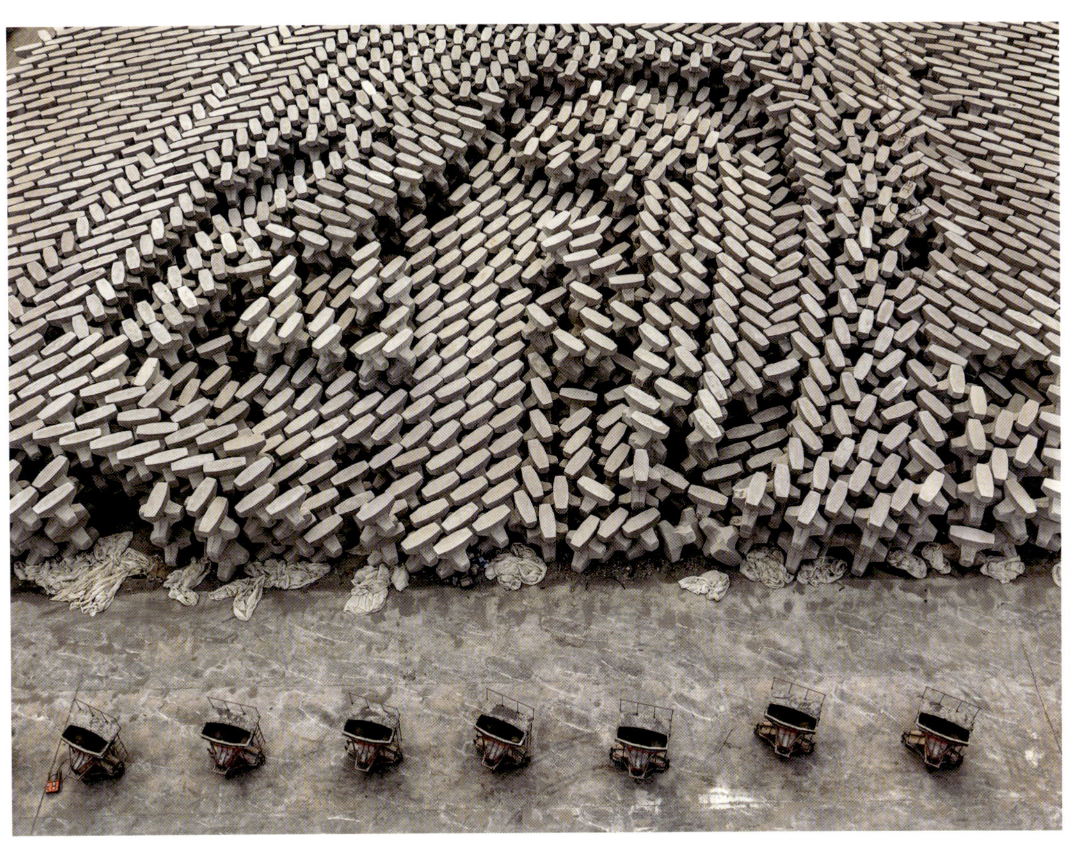

Edward Burtynsky, *Tetrapods #1, Dongying, China*, 2016

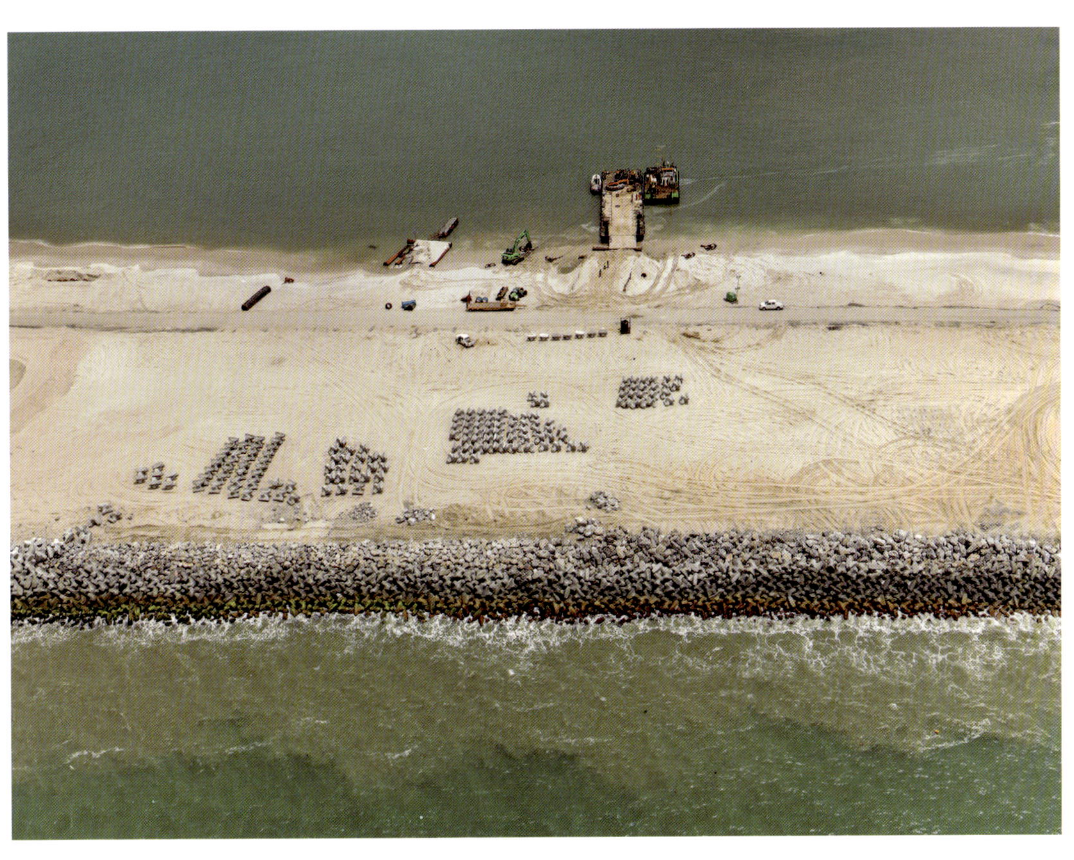

Edward Burtynsky, *Eko Atlantic Development #1, Lagos, Nigeria*, 2016

Over the past decade or two, Lagos has emerged as one of the economic capitals of Western Africa. As the city is composed of a series of islands and peninsulas on the Gulf of Guinea, natural processes of coastal erosion are exacerbated by climate change, putting much of Lagos at a high risk of flooding. With land at a premium, low-income residents and private developers have moved to create their own real estate. Since 1970, Lagos has grown at breakneck speed, from a city of around 1.4 million inhabitants to one of more than 20 million people in just two generations.[1] The wealth disparity is stark, and a significant number of residents live in dense informal settlements.[2] Makoko, and communities like it, have no fortifications to protect inhabitants from climate change. Built on the water without infrastructure, the informal settlement boasts a population estimated to range from 40,000 to 300,000 people.[3] Eko Atlantic, a nearby multi-billion-dollar artificial peninsula, has plans to house a quarter of a million people. Over ten square kilometres of sand were dredged to create the peninsula that is attached to Victoria Island, Lagos' financial centre.[4] With the peninsula highly vulnerable to storm surges, the project includes the construction of a seawall known as the "Great Wall of Lagos." Approximately 100,000 five-ton concrete blocks will be arranged to form the wall, which will extend over eight kilometres.[5] However, what might protect the wealthy districts of Eko Atlantic and Victoria Island may end up directing damage from a storm to unprotected informal communities. Eko Atlantic, designed to employ 200,000 commuters largely in the financial and tech sectors, is meant to represent the new, wealthy Lagos. The real estate, including nearby Bar Beach, is so valuable that, in 2008, local police conducted violent economic evictions in the run-up to construction.[6] Nigeria is the most populous country in Africa, with a rapidly growing population that by 2018 had expanded to more than 190 million people. In 2017, the United Nation's World Population Prospects found that among the ten largest countries in the world, Nigeria's population was growing the most rapidly.[7] As such, the population of Nigeria is projected to surpass that of the United States shortly before 2050, at which point it would become the world's third-largest country by population.[8] Integral to the city's fabric, the growing low-income communities of Lagos are nonetheless precarious in the face of climate change, politics, and capital.

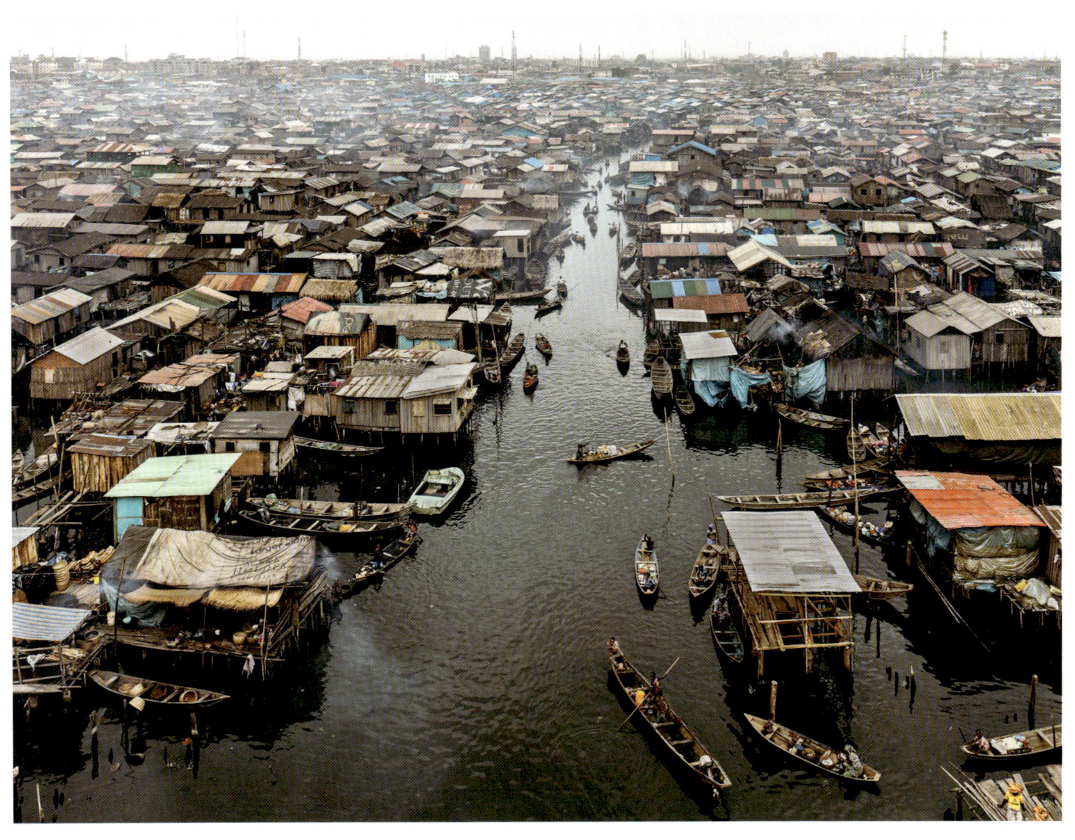

Edward Burtynsky, *Makoko #2, Lagos, Nigeria,* 2016

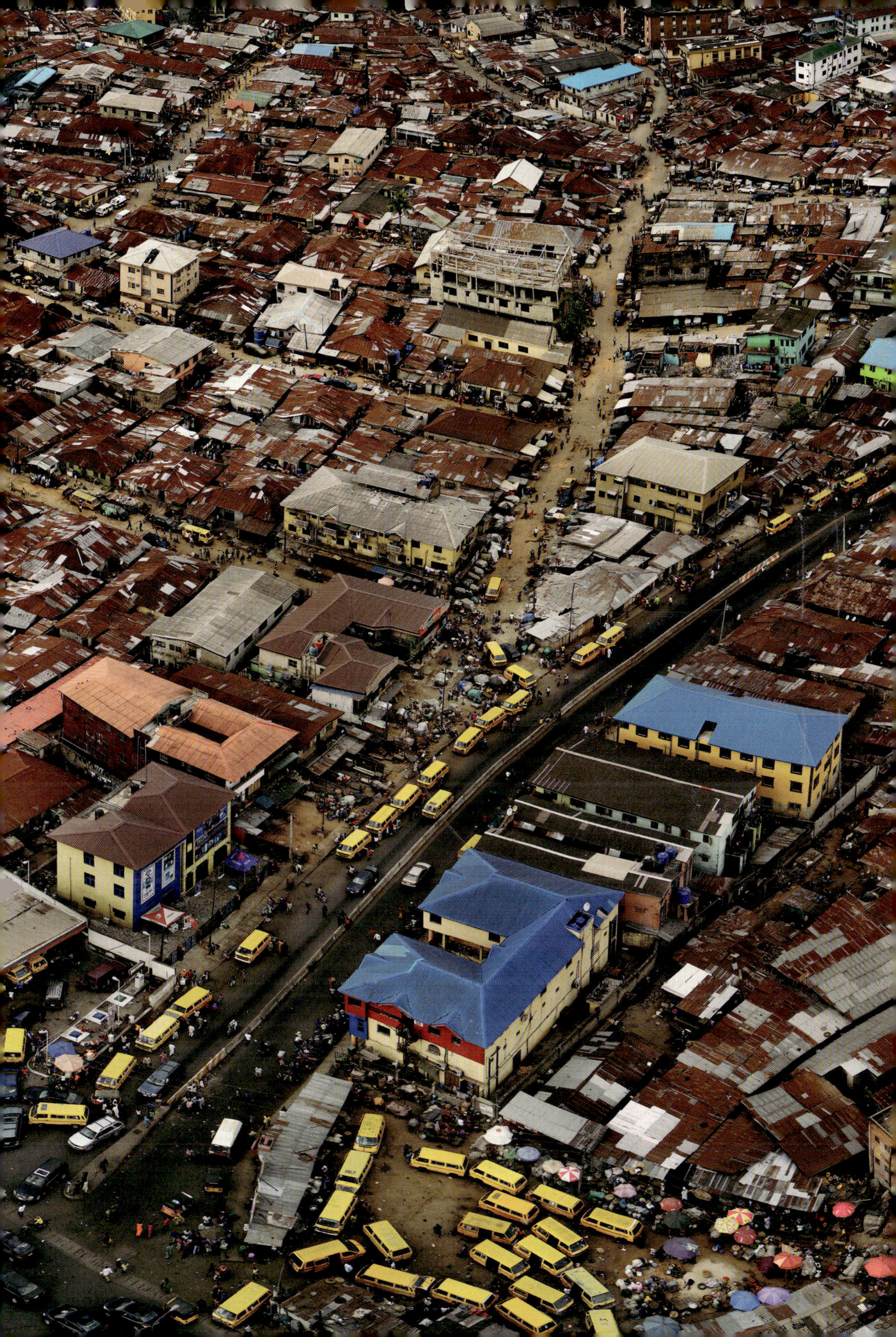

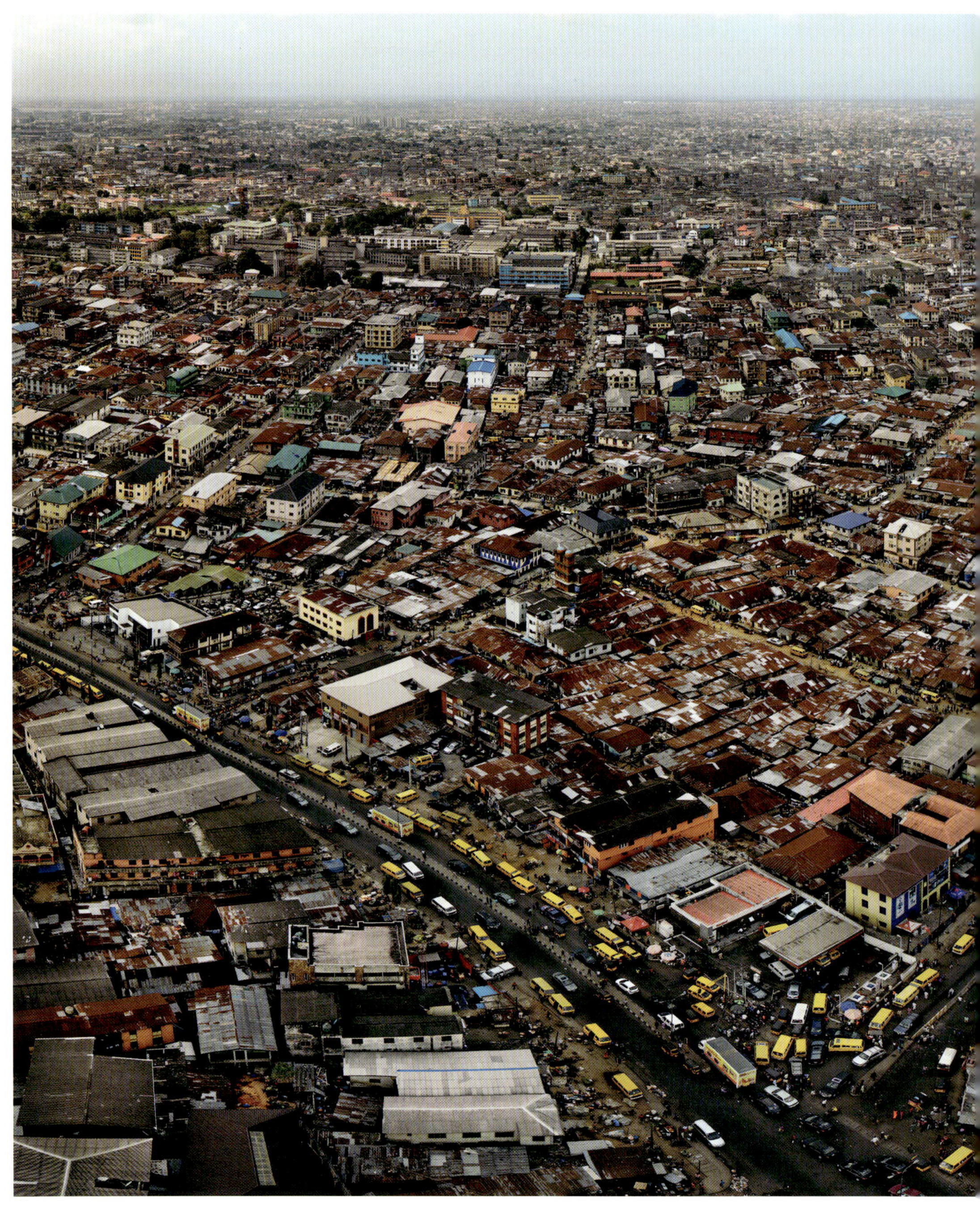

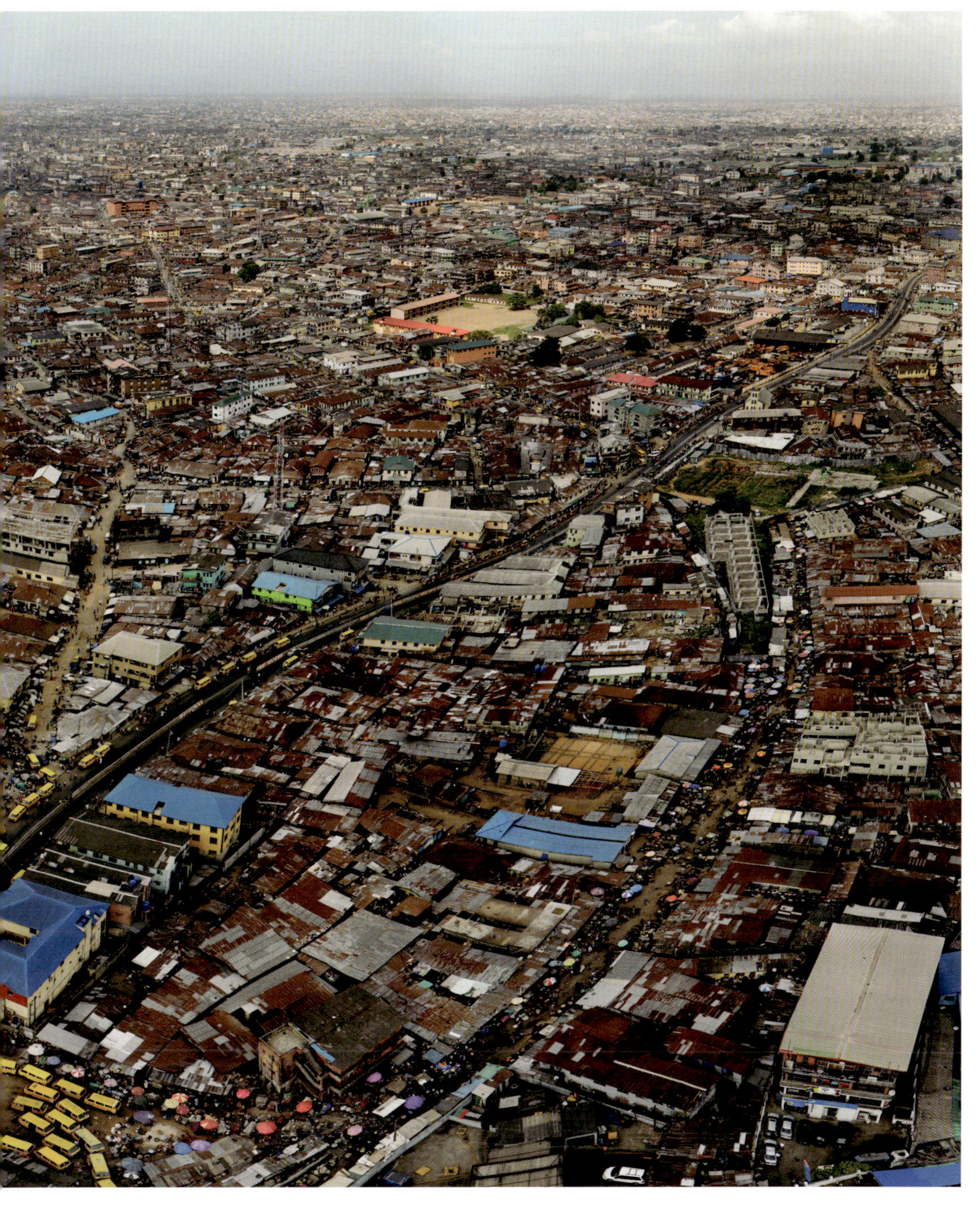

Edward Burtynsky, *Mushin Market Intersection, Lagos, Nigeria*, 2016–2018

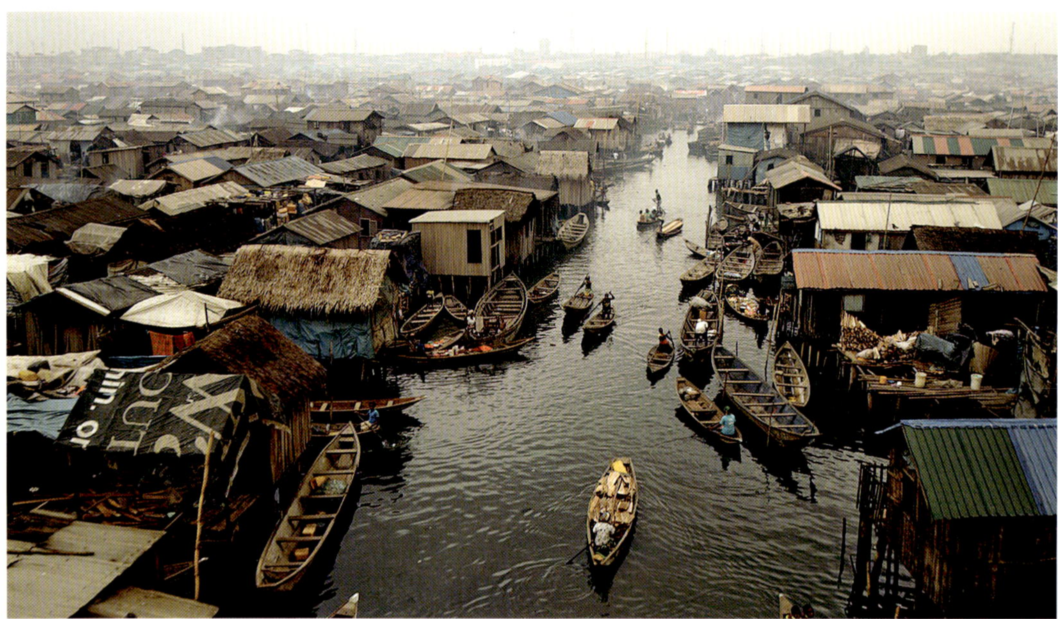

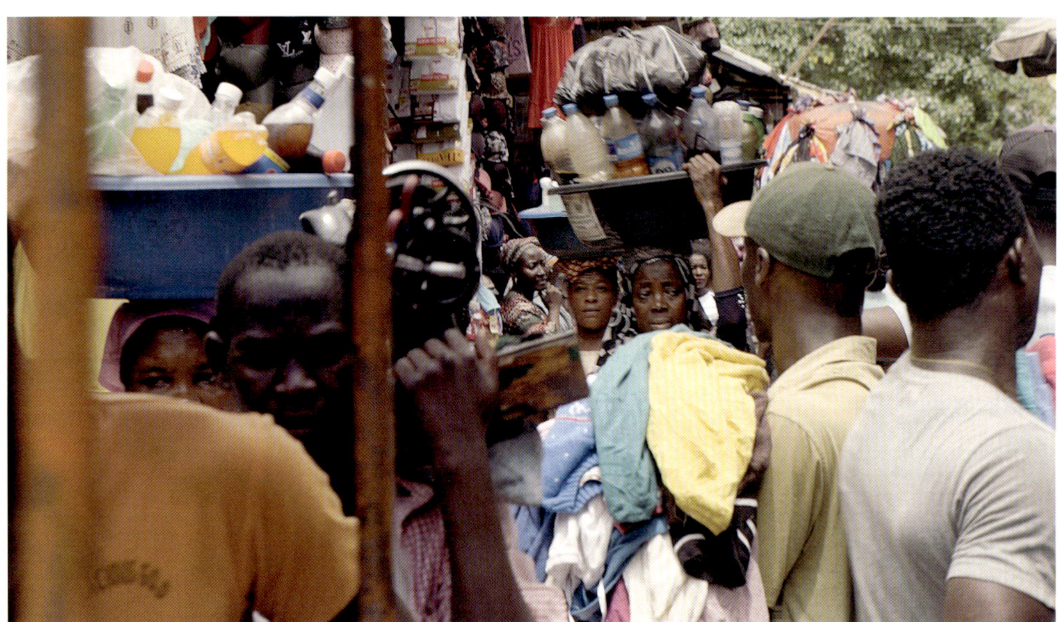

Jennifer Baichwal and Nicholas de Pencier
Makoko, Lagos, Nigeria, 2018
Market Walks, Lagos, Nigeria, 2018

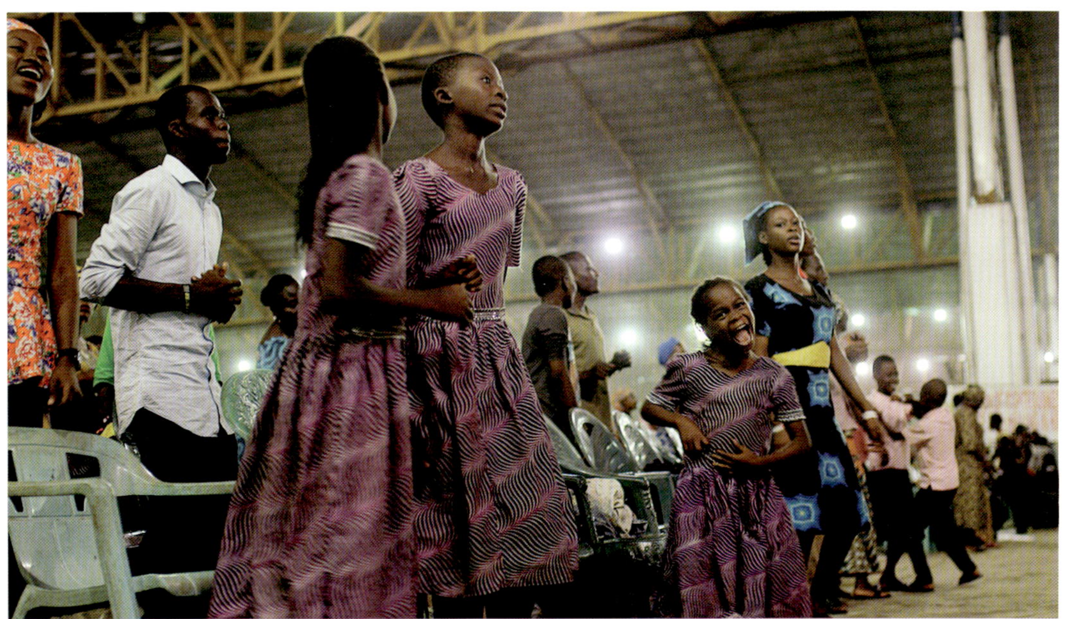

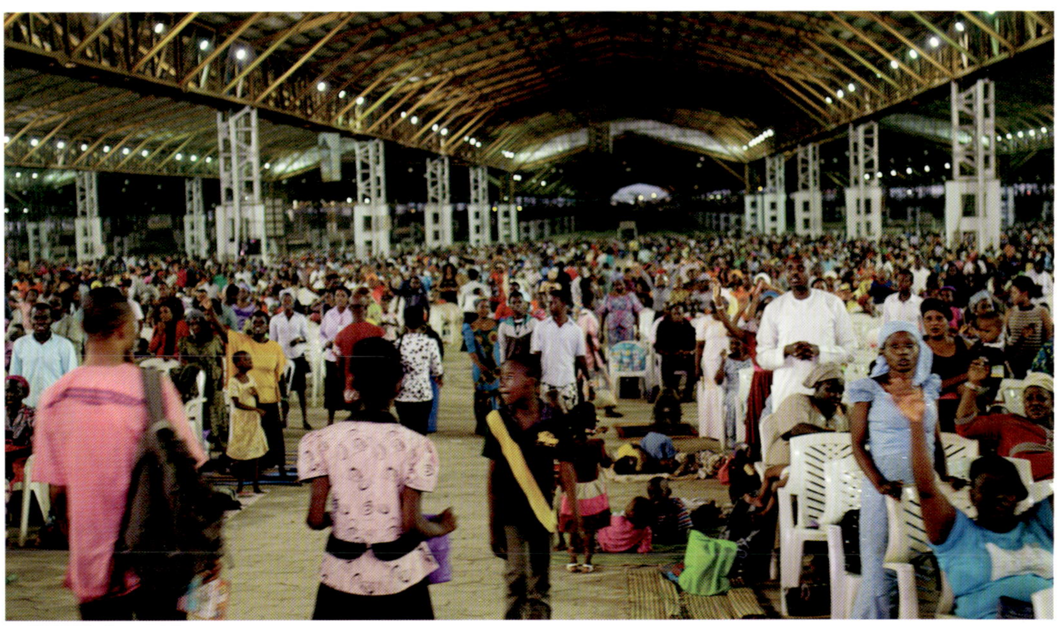

Jennifer Baichwal and Nicholas de Pencier
Redeemed Christian Church of God, Lagos, Nigeria, 2018

Urban sprawl is not easily described, yet most people seem to know it when they see it. Though it has become increasingly dense over time, Los Angeles represents the origin of and model for the type of urban sprawl we see today: large areas of separated land use, single family houses with yards, and a reliance on the automobile. It is the archetypal image of the American Dream, exported around the world. The development of the California freeway system in the early to mid-twentieth century was a catalytic moment for the development of large-scale urban sprawl. In the twenty-first century, the explosion of urban development in China has outpaced the rate of change elsewhere. As with the early development of urban sprawl in the Los Angeles area, the critical infrastructure required for this type of expansion in China is the highway. Much like California's suburbs, China's new megacities are marked by low-density development, which presents numerous footprint issues as the global population grows.[1]

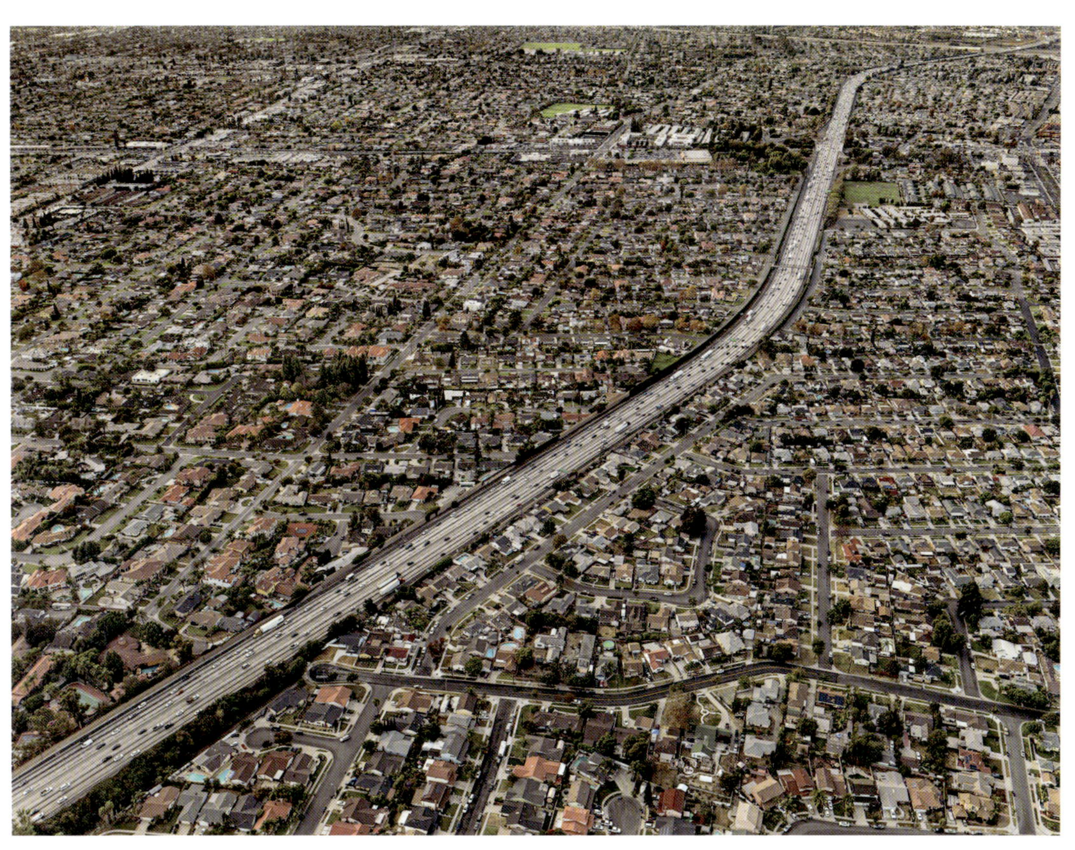

Edward Burtynsky, *Highway #8, Santa Ana Freeway, Los Angeles, California, USA*, 2017

In the fall of 2017, between thirty and forty thousand cars accumulated at the Royal Purple Raceway in Baytown, Texas, as a result of Hurricane Harvey.[1] Cutting across major population areas in Texas and Louisiana, the fatal tropical storm caused significant property damage to homes and businesses. Estimates put the total losses from the storm at as much as $125 billion USD, according to NOAA (National Oceanic and Atmospheric Administration). These inoperable, storm-damaged cars were temporarily stored at this site before eventually being returned to their owners, sold, or designated for demolition.[2] The cars pictured here are most likely insurance write-offs. Within view of this yard are the sprawling industrial complexes of Houston's petroleum industry (page 159). Over 45 percent of the U.S. petroleum refining capacity, as well as 51 percent of total U.S. Natural Gas processing plant capacity, is located along America's Gulf Coast.[3] In total, federal offshore oil production in the region accounts for 17 percent of U.S. crude oil production.[4] During the hurricane, 11 percent of the total refining capacity of the U.S. was reduced as refineries were forced to shut down, and a quarter of American oil production from the Gulf was halted.[5] Damage from the storm caused significant leakage of harmful chemicals, including carcinogens, into the air and waterways.[6] Petroleum products are the key component of the asphalt that coats our roadways, the fuels and lubricants that power our vehicles, synthetic fertilizers that farmers use to bring in high-yield crops, and the plastics that package our goods. As such, the sum total of its production over history has directly caused the increasing CO_2 levels that are now changing our climate. A one-degree Celsius increase in global average temperature would cause extreme weather events, like Hurricanes Harvey and Katrina, to occur with seven times the frequency.[7] While these catastrophic events are easily forgotten by those who are unaffected, once the clouds have passed, their cumulative impact on local communities will only increase alongside global carbon levels.

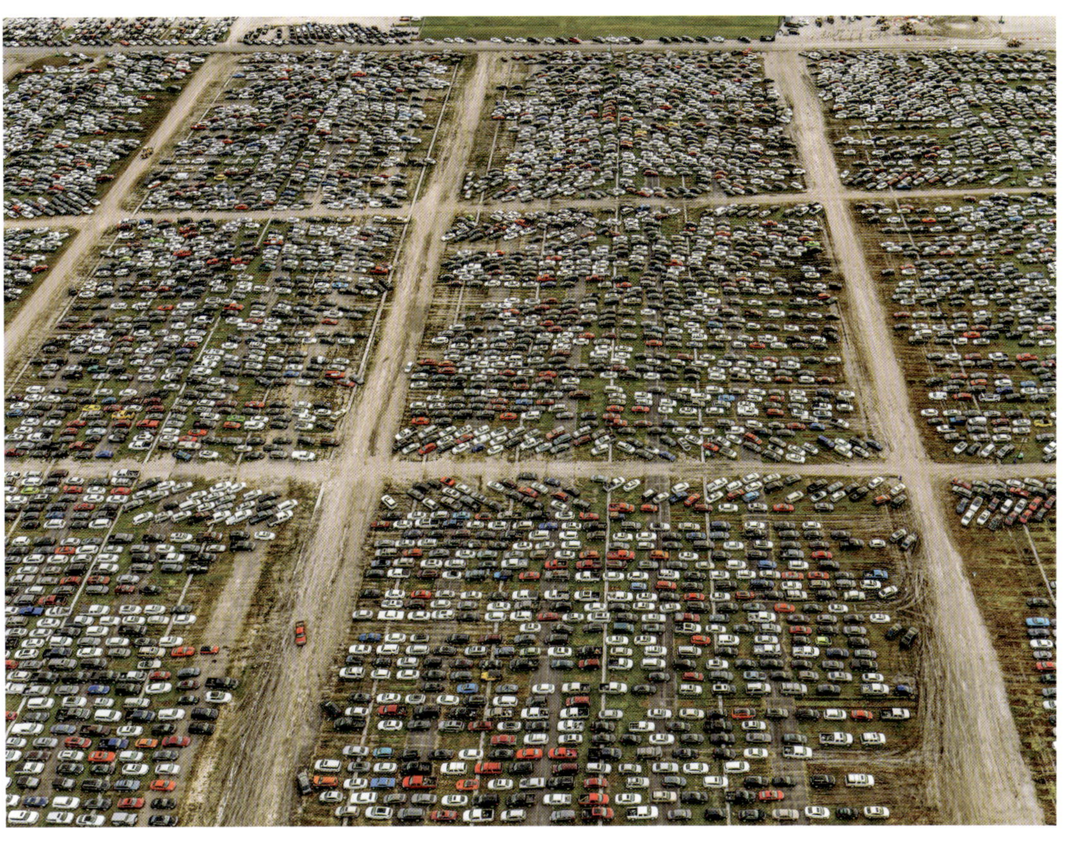

Edward Burtynsky, *Flood Damaged Cars, Royal Purple Raceway, Baytown, Texas, USA*, 2017

The coastal marshes of Spain's Bay of Cádiz are prime examples of human-engineered biodiversity. Initially developed as salt collection ponds, polyculture fisheries were created as a means of expanding food production. A pond and marsh system, fed by two rivers as well as the tide, create this dramatic landscape. As with solar salterns, a series of ponds are built into naturally occurring salt marshes and connected by channels, encouraging a gradient of water depth and salinity. The yearly farming cycle begins in winter, when the floodgates of the reservoirs are opened to empty the pools and harvest the fish produced during the previous season.[1] The gates then remain open for several months throughout winter and spring, before being shut once again, trapping fry (newly hatched fish) along with the organisms that will feed them.[2] The resulting fully grown fish, as well as other species such as molluscs and crustaceans, are then collected in the autumn and early winter.[3] This traditional system has always encouraged experimentation and adaptability. By the mid-twentieth century, the salt industry had seen a serious decline, and fish farming was increased to compensate. By the 1990s, advances in technology allowed for production to be carried out on an industrial scale. Today, traditional fish farms, while more sustainable, are not as profitable as the rapidly expanding monoculture operations. The Bay of Cádiz is considered a natural park and heritage zone, yet many of the traditional ponds have been left abandoned. It may be possible for polyculture to survive alongside industrial fish farms, but without maintenance, this landscape and the biodiversity it fosters may disappear through erosion and drying out by sedimentation.[4]

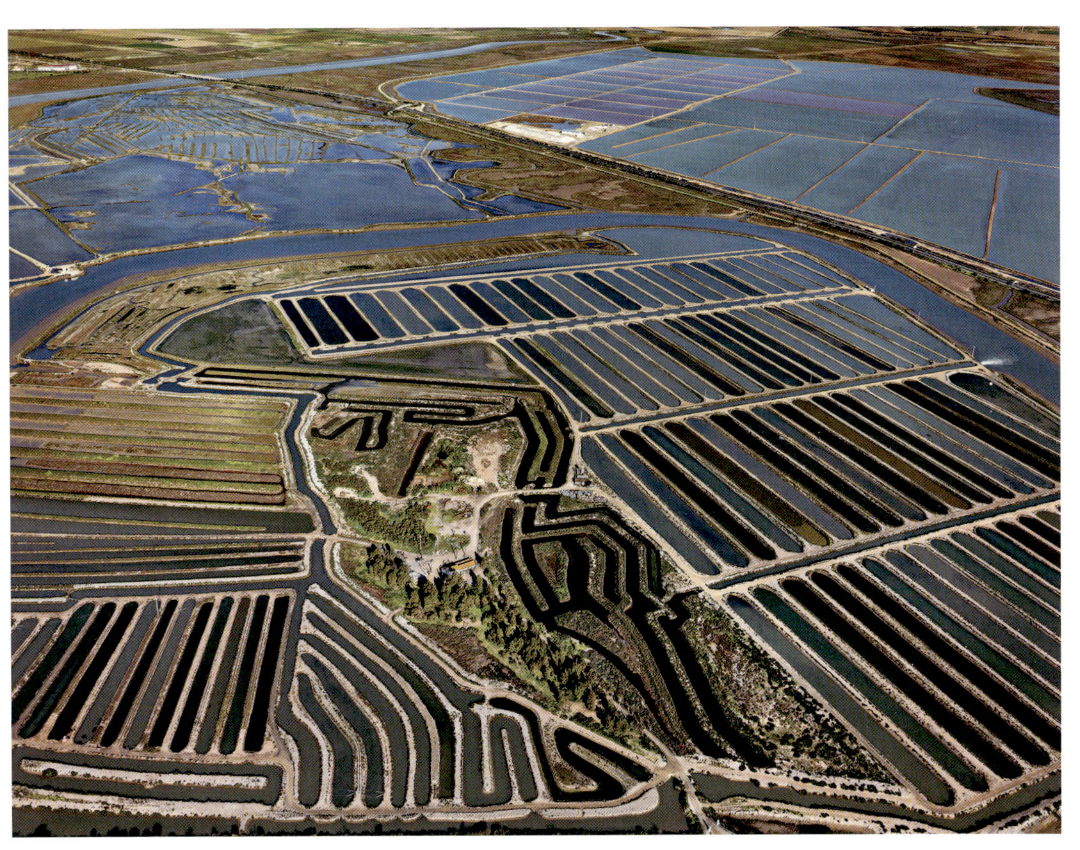

Edward Burtynsky, *Salinas #5, Aquaculture, Cádiz, Spain,* 2013

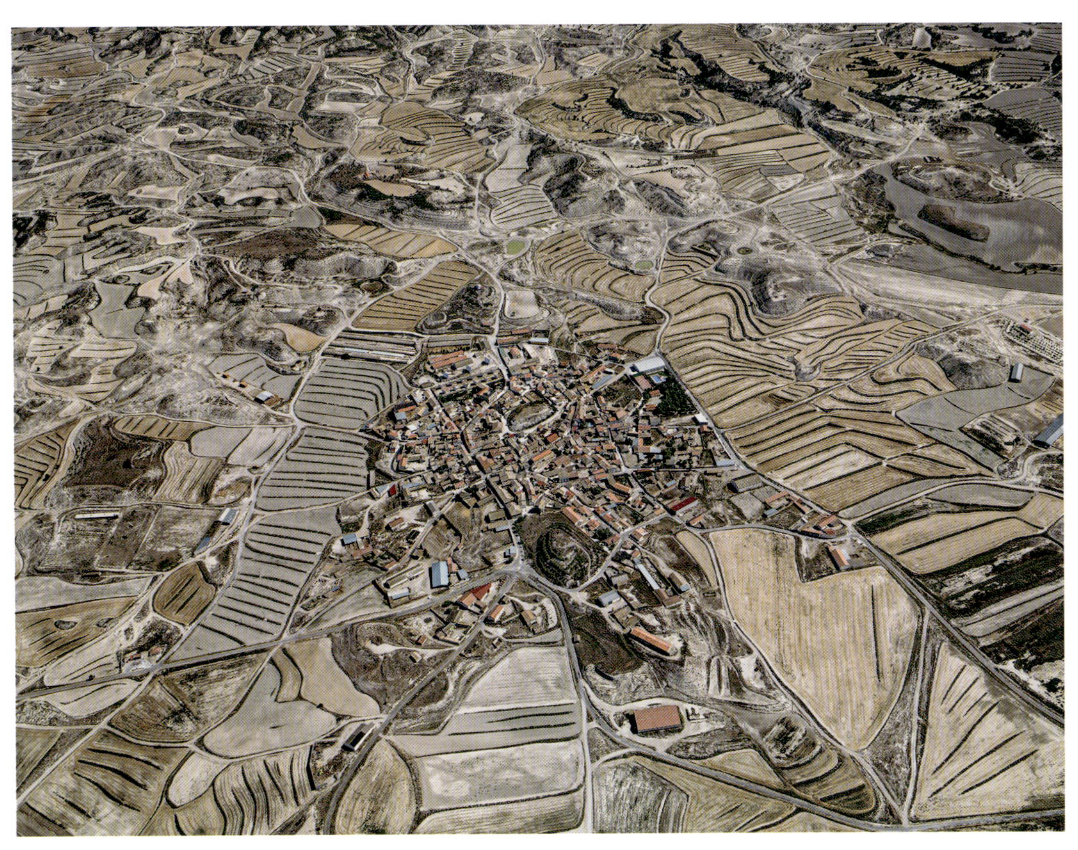

Edward Burtynsky, *Dryland Farming #40, Monegrillo, Spain,* 2010

The Imperial Valley is a low-lying desert located at the southeast end of the Salton Sea. Since the 1940s, when significant water resources were made available through the All-American Canal, the Imperial Valley has become an important agricultural region, growing $2.3 billion worth of agricultural products in 2013,[1] most famously, winter salad vegetables. As the valley's only water source, the canal controls about 20 percent of the Colorado River's total annual output.[2] Agricultural runoff drains directly into the nearby sea, carrying salts from the earth as well as nitrates and phosphorus, thus encouraging the growth of algae and bacteria, which further rob the sea of oxygen. Today, more than 80 percent of consumptive water use in the United States is for agriculture.[3] On average, Imperial County agriculture uses 5.6 acre-feet of water per acre per year,[4] putting significant pressure on the Colorado River water supply. In recent years attempts have been made to reduce water consumption. Several programs pay farmers annual per-acre fees to leave their land fallow. Many hope that this system will motivate farmers to conserve water even further.[5] However, while fertilizer runoff has encouraged bacteria and algae growth, and significantly increased the salinity of the Salton Sea, reductions in Imperial Valley irrigation may well have their own negative impacts on the rapidly evaporating lake. The historical tension between urban and rural livelihoods continues as farmers, who use 85 percent of Colorado's water supply,[6] struggle to change their practices.

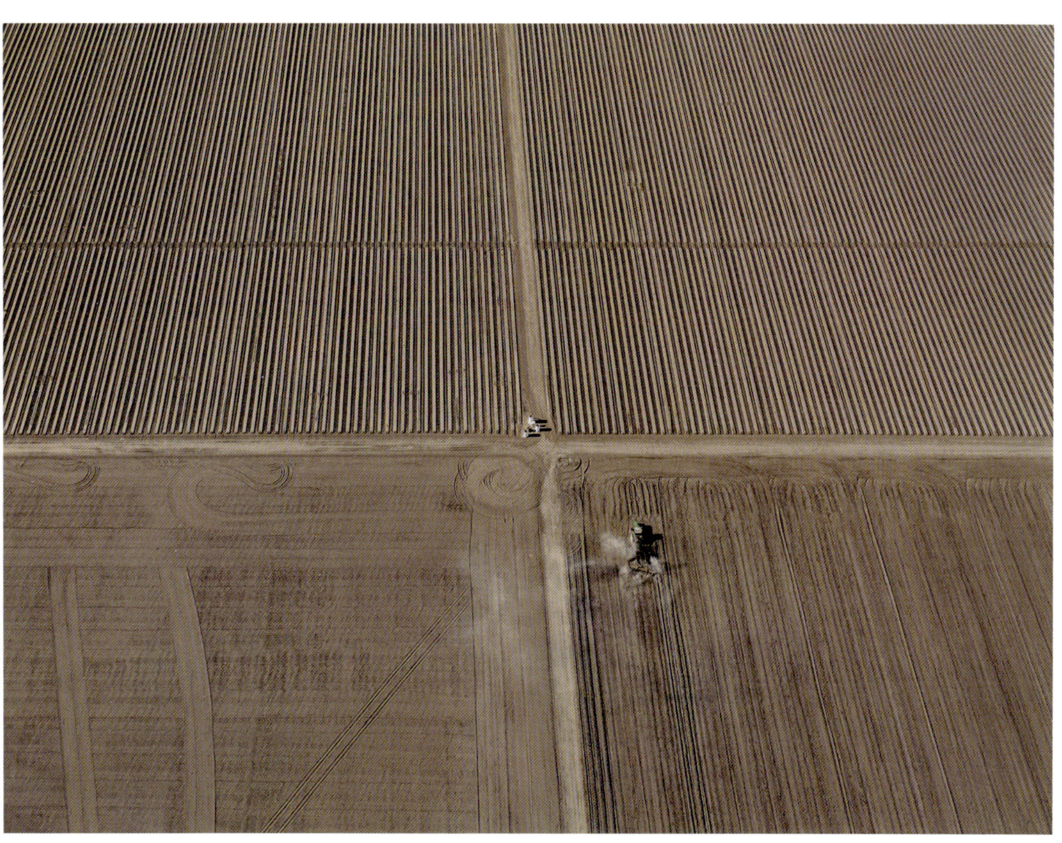

Edward Burtynsky, *Imperial Valley #4, California, USA*, 2009

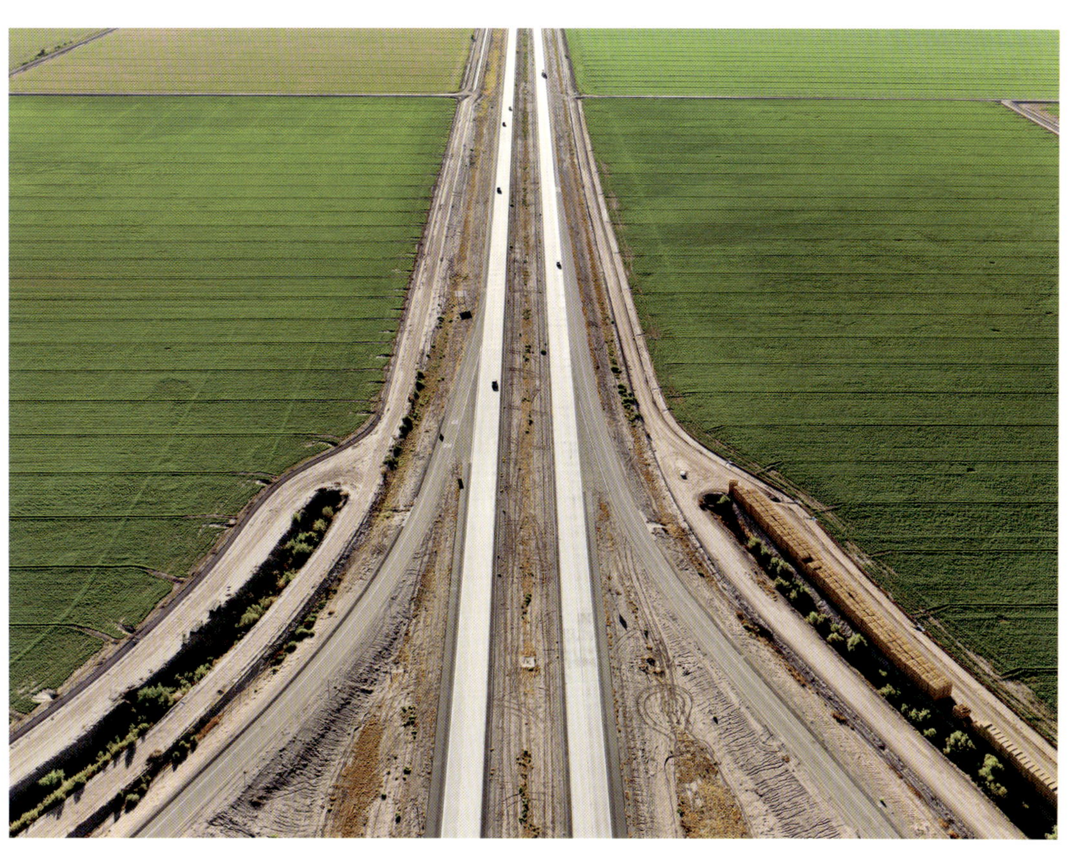

Edward Burtynsky, *Imperial Valley #5, Holtville, California, USA,* 2009

Since the nineteenth century, farmers in many arid regions of Spain have practised dryland agriculture. Using terraforming techniques, combined with crops resistant to dry conditions, these farmers have been able to make barren land productive. Yet the introduction of anthropogenic materials in the period following the Spanish Civil War has caused another type of earth-shaping for agriculture in the region. The province of Almería in southern Spain has come to be known as the "Mar de plástico" ("Sea of Plastic").[1] The province's industrialization of vegetable production has transformed the community, once one of the poorest regions in the country, into one of the richest in Spain.[2] In order to provide fresh vegetables for European cities, almost four hundred square kilometres around the municipality of El Ejido are today completely covered with plastic greenhouses.[3] Additionally, the workforce is almost entirely made up of migrant workers, mostly from Africa, working in conditions that a 2011 report in the *Guardian* considered to embody the UN's definition of "modern-day slavery."[4] Their inhumane living conditions have been compared to those of a refugee camp. Further, the local Sorbas-Tabernas fossil water aquifer, an ancient and non-renewable resource, is now in danger of collapse. The situation was fictionalized in the 2015 Spanish television crime drama *Mar de plástico,* which received condemnation from industry representatives.[5] Yet problems continue to be documented by journalists and human rights groups. As global demand for cheap, healthy diet staples increases, it remains vital that the human cost of these products is not obscured from their consumption.

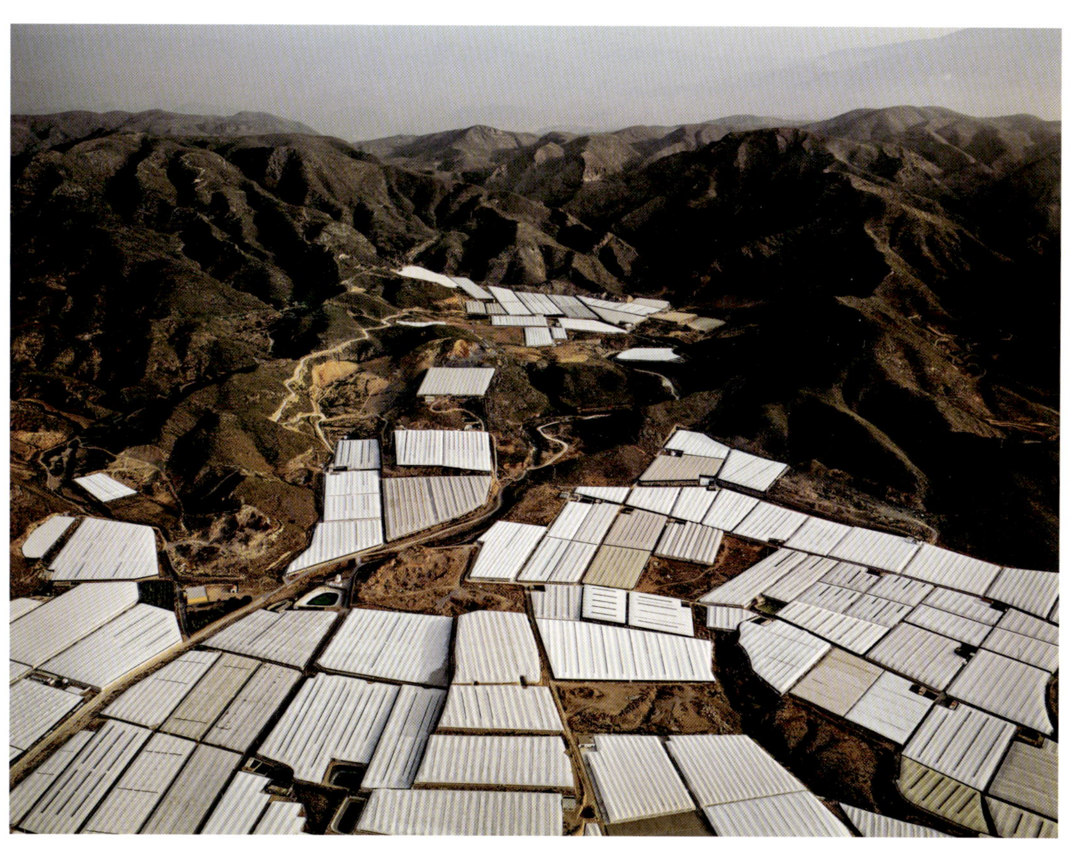

Edward Burtynsky, *Greenhouses #2, El Ejido, Southern Spain,* 2010

Agricultural areas in the Saudi Peninsula are irrigated with what is known as "fossil water"—water that was trapped in underground aquifers during the last Ice Age.[1] As this water source is a non-renewable resource, it is estimated by NASA that these aquifers will run dry within fifty years.[2] The area that's pictured in this satellite image is about 163 square kilometres. It is estimated that a single pivot can cover an area from 0.4 square kilometres to 0.9 square kilometres.[3] The pumps require around 91 litres of diesel fuel per hour—another non-renewable resource made accessible because of significant fuel subsidies.[4]

Pivot irrigation, developed in the United States, was first used in Saudi Arabia in the late 1970s for water-intensive wheat crops. Reputed to be more than twice as efficient as flood irrigation, pivot irrigation—designed to irrigate crops in circles—allowed harvest yields to increase and irrigated areas to rapidly expand. By the early 1990s, government programs to intensify production resulted in Saudi Arabia becoming a top global exporter, putting greater stress on water consumption.[5] By the time this image was captured, production had switched to higher-value animal feed such as alfalfa and corn. Yet decreasing water levels had reduced the total farmed area by more than half since the 1990s. At the same time, spring-fed oases, existing since biblical times, dried out as underground water sources were consumed by agriculture.[6]

Most operations today use more water-efficient drip irrigation. Israeli technology, for example, delivers the exact amount of water that is needed to the roots of each plant—with no evaporation or runoff. However, these systems are very expensive, and adoption can be slow without serious incentives.[7] As of 2017, 160 square kilometres of land have been converted to more sustainable organic farming practices, representing only 2 percent of the kingdom's total agricultural area. Such changes must occur rapidly if the imminent collapse of Saudi agriculture is to be avoided.[8]

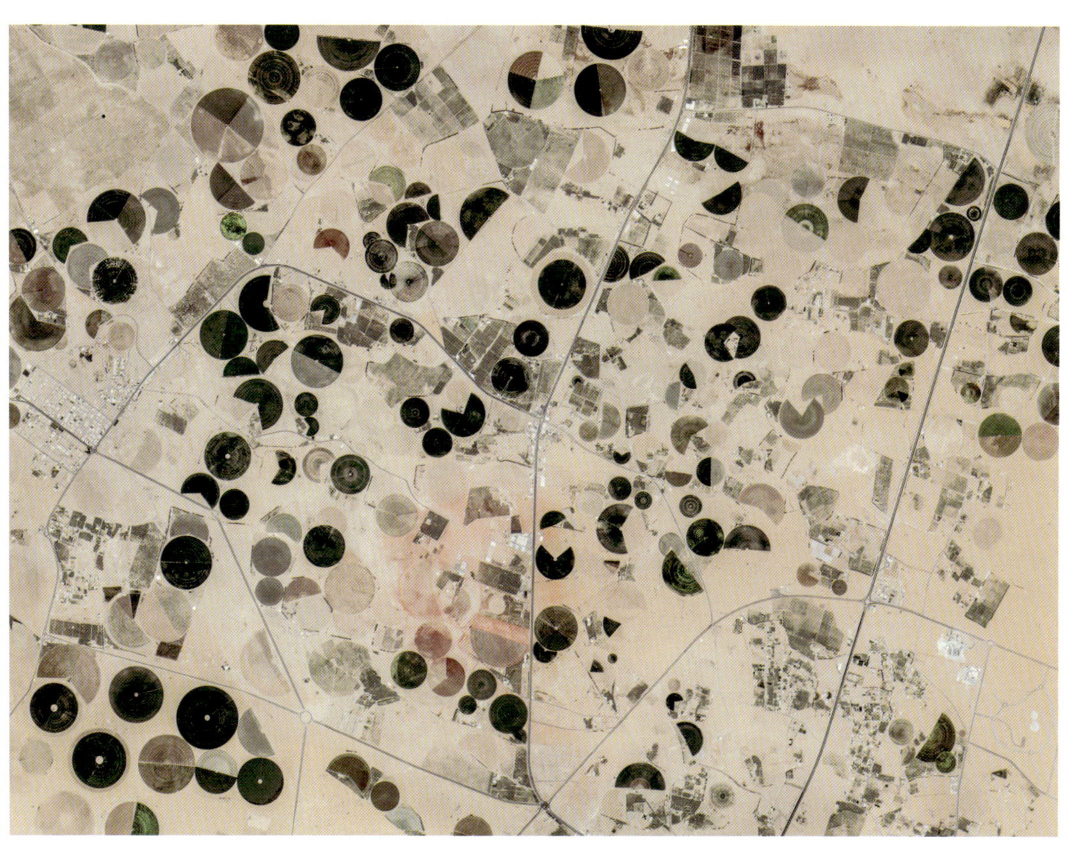

Edward Burtynsky, *Satellite Capture, Near Buraydah, Saudi Arabia,* 2018

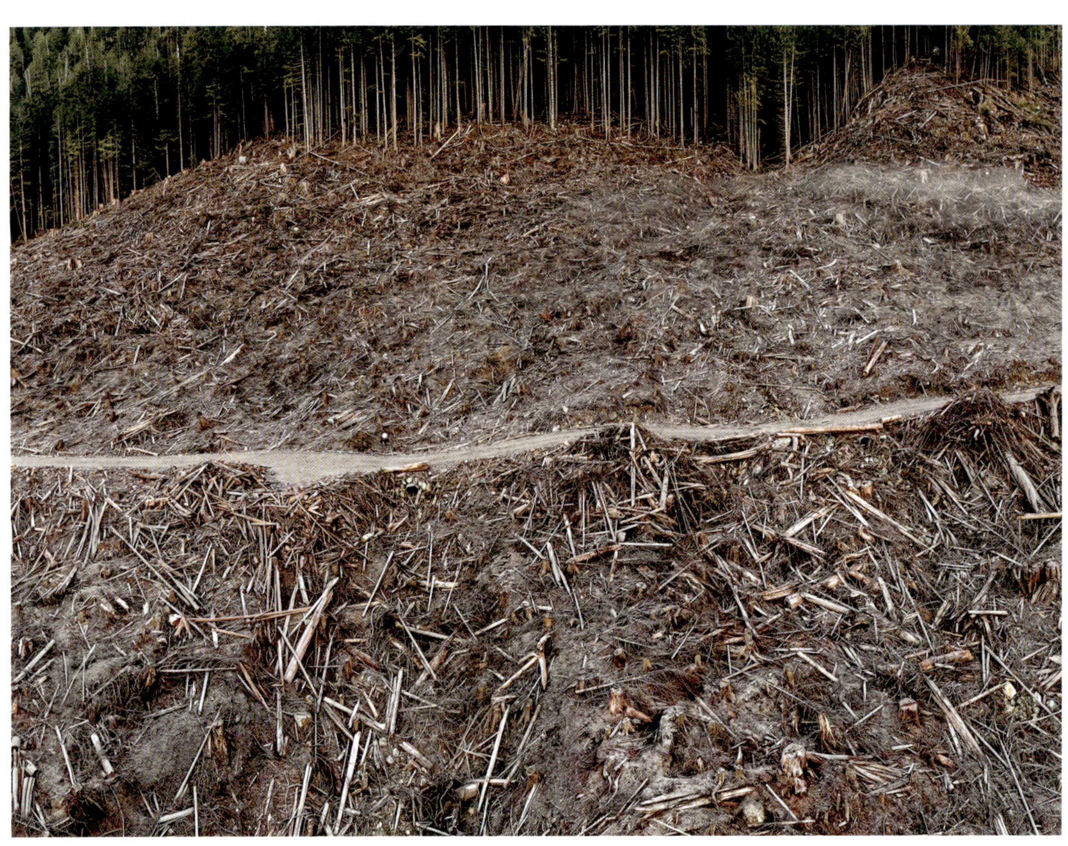

 Edward Burtynsky, *Clearcut #5, Vancouver Island, British Columbia, Canada,* 2017

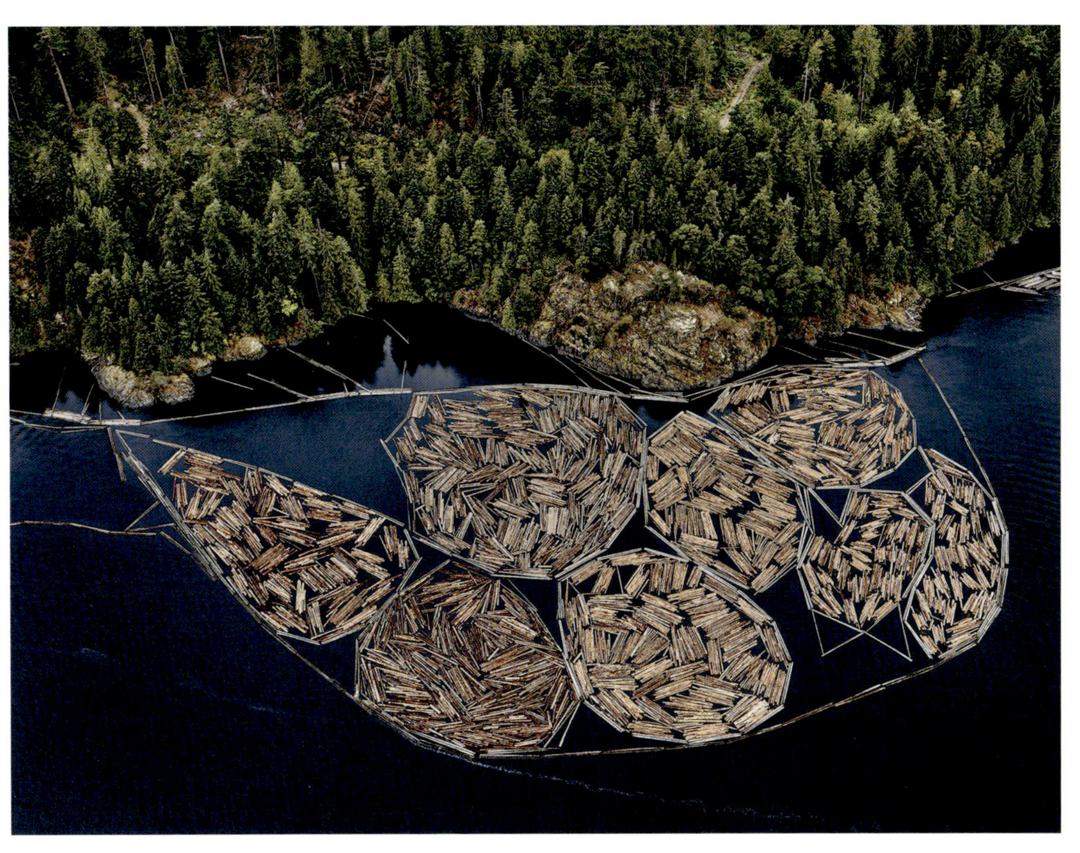

Edward Burtynsky, *Log Booms #1, Vancouver Island, British Columbia, Canada,* 2016

In 2017, *The Anthropocene Project* team took three trips to Port Renfrew, on the south end of Canada's Vancouver Island, to photograph the second-tallest Douglas fir in the country. The tree, which is believed to be about one thousand years old, is approximately sixty-six metres high, which is nearly the height of a twenty-storey building.[1] The tree stands alone in an otherwise clear-cut area of the forest, and has been aptly named "Big Lonely Doug." The old-growth trees that once surrounded Big Lonely Doug have been logged, leaving it vulnerable and at risk of being blown down by wind. The number of large old-growth trees is declining rapidly as the timber industry fells the last of these giants. As awareness of Big Lonely Doug has spread, though, thanks to local activism, tourism to the area and related public pressure to protect the old-growth forests have increased.

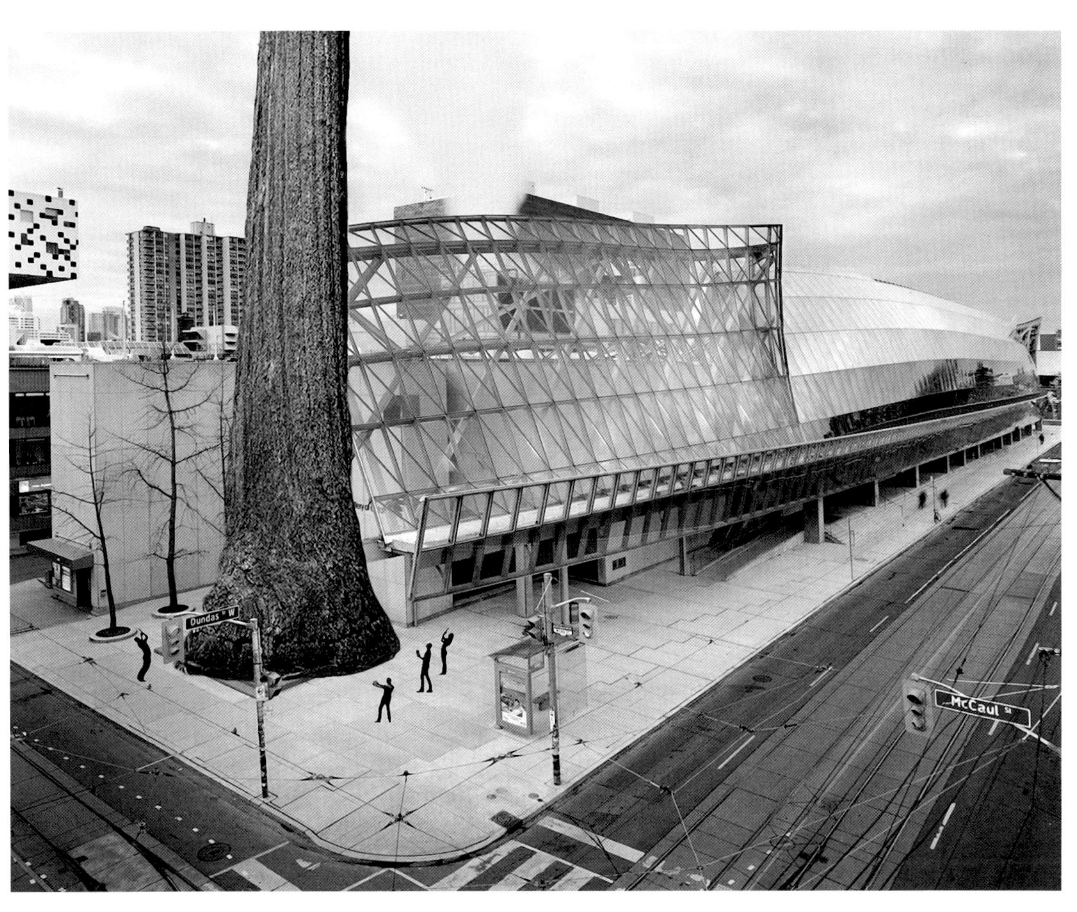

Edward Burtynsky, Jennifer Baichwal, and Nicholas de Pencier,
AR #3, Big Lonely Doug, Vancouver Island, British Columbia, Canada, 2016

Starting in August 2015, fires consumed huge areas of tropical rainforest across Indonesia, a catastrophic event considered to be one of the worst environmental disasters of the century. By October, almost 16 million tons of CO_2 were being emitted each day—greater than the daily emissions generated by the entire U.S. economy.[1] Borneo, an island politically divided between three countries—Indonesia, Brunei, and Malaysia—was heavily impacted. The fires were lit as part of a seasonal phenomenon in which land is set ablaze in order to repurpose it at low cost.[2] As much of the affected area is composed of highly flammable peatland, initially, small scale fires spread at a rapid and ferocious rate. This practice is commonly referred to as "slash and burn," as the newly cleared land can then be used for industrial palm oil production.[3] A palm plantation takes three to four years after planting to bear fruit,[4] and each palm is productive for 25 to 30 years.[5] At this point, the plantation is often burned and the cycle starts again. It is estimated that between 1990 and 2005, 55 to 60 percent of palm plantations in Indonesia and Malaysia were planted on former virgin tropical forest.[6] Between 2011 and 2013, 40 percent of the total deforestation across the island was for palm, and a fifth of the deforestation occurred in areas designated for a moratorium on plantations.[7] This trend is not new. Between 1973 and 2010, almost a third of Borneo's forests disappeared.[8] Despite government designations for protected forests, and corporate "zero-deforestation" policies, the directives are unenforced. Thus, these areas continue to be logged and converted to plantations.[9] Oil palms require significant amounts of nitrogen to grow and potash to produce fruits;[10] the resulting application of fertilizers is the most polluting element in the agricultural phase of palm oil production.[11]

This intensive form of agriculture's near-ceaseless expansion is a direct result of the global demand for cheap food-grade oils such as palm, of which Indonesia and Malaysia produce 85 percent of the world's supply.[12] One of the world's top commodities, palm oil is found in everyday household products like toothpaste and processed foods. As a result, there have been significant efforts from grassroots organizations to begin working with select corporations on sustainable palm oil production practices. Currently, smallholders (i.e., independent local farmers) produce over 40 percent of the world's supply.[13] Cycling through their already developed agricultural land (rather than recently cleared forest) and using the chipped stems of older palms to return nutrients to the soil (rather than burning) are some of the techniques being used to reduce current burning and forest clearances. Certification systems, such as those advocated by the Palm Oil Innovation Group (POIG, a coalition developed in partnership with leading NGOs as well as with progressive palm oil producers), have been shown to have some success in reducing the amount of exploitative palm entering the market.[14] Yet without the full cooperation of the large multinationals that purchase the refined palm oil, these systems cannot altogether halt the large-scale destruction of these peatlands. Without sustained effort and cooperation, the complexities of the global supply chain will encourage the continued exploitation of tropical land and peoples, and contribute to the rapid shrinking of one of the orangutan's last remaining habitats.

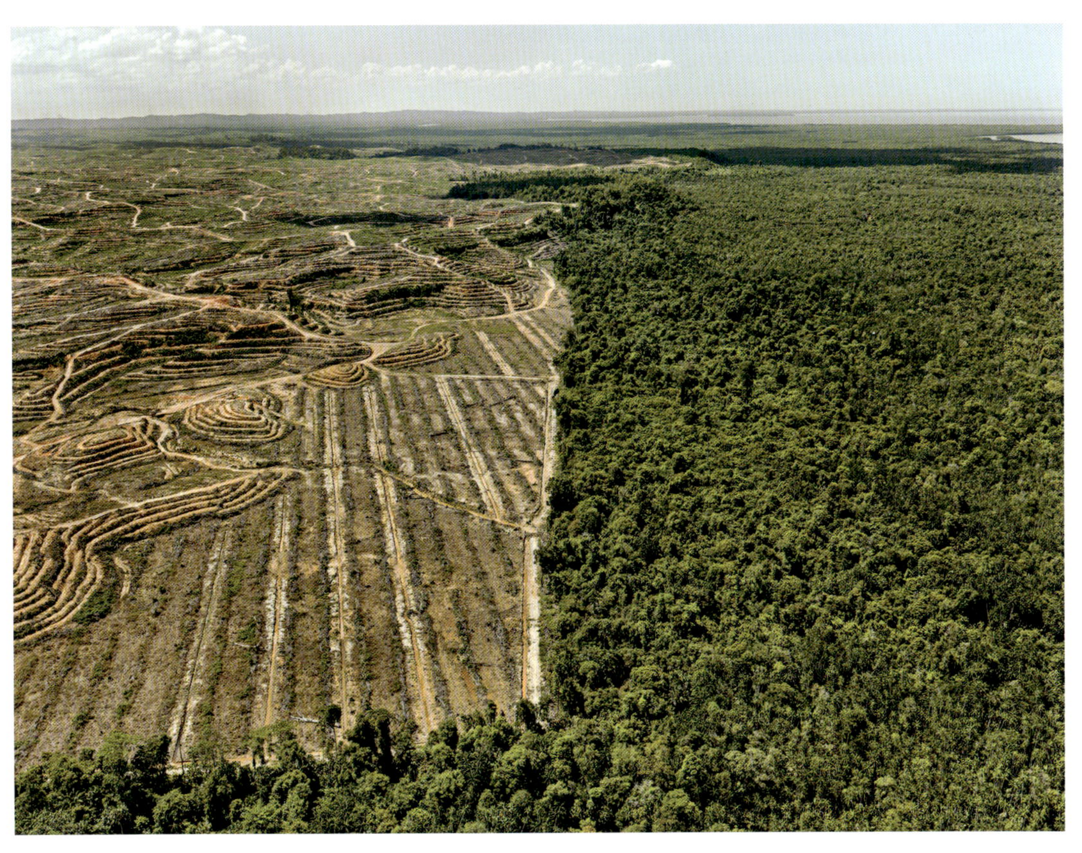

Edward Burtynsky, *Clearcut #1, Palm Oil Plantation, Borneo, Malaysia*, 2016

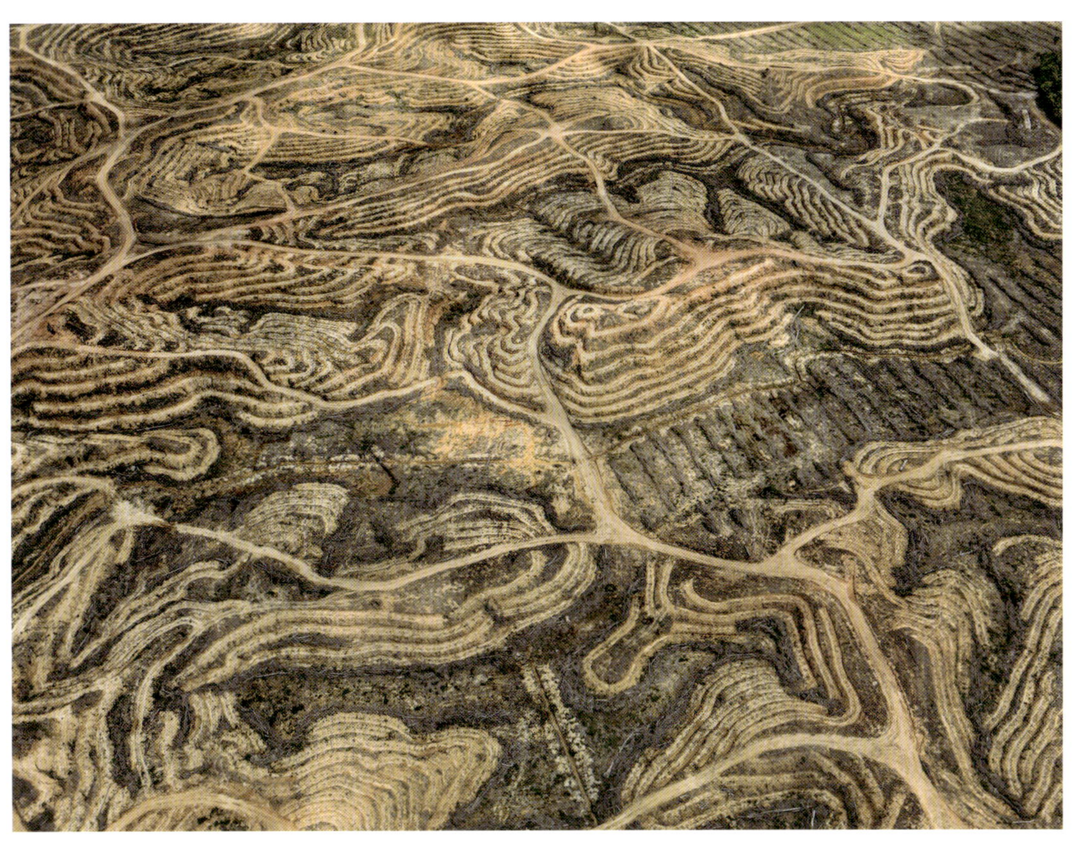

Edward Burtynsky, *Clearcut #3, Palm Oil Plantation, Borneo, Malaysia,* 2016

As one of the fastest-growing countries in the world, Nigeria has seen large areas of its ancient forest decimated, especially in recent years. Log booms, such as the ones pictured here, transport raw timber downriver from the Niger Delta to Makoko, the final destination for much of the collected resource. The densely populated, informal settlement is home to an expansive forest industry, with abundant sawmills and myriad markets for wood. As Makoko is situated atop the water, thus earning its nickname, "Venice of Africa," the logs that reach Makoko's shores can be easily accessed, processed, and sold to eager markets.

The emergence of Nigeria as an increasingly industrial nation has not been without cost: by 2011, 40 percent of the Delta's lowland rainforests had been lost due to agriculture and logging, while freshwater forests had been reduced by nearly a third.[1] Over time, biodiversity in the region has steadily diminished.[2] For the wood merchants of Makoko, the threat of this increasing deforestation is tangible—their primary source of income is mostly dependent on a natural resource in such high demand that its popularity will inevitably result in its disappearance.

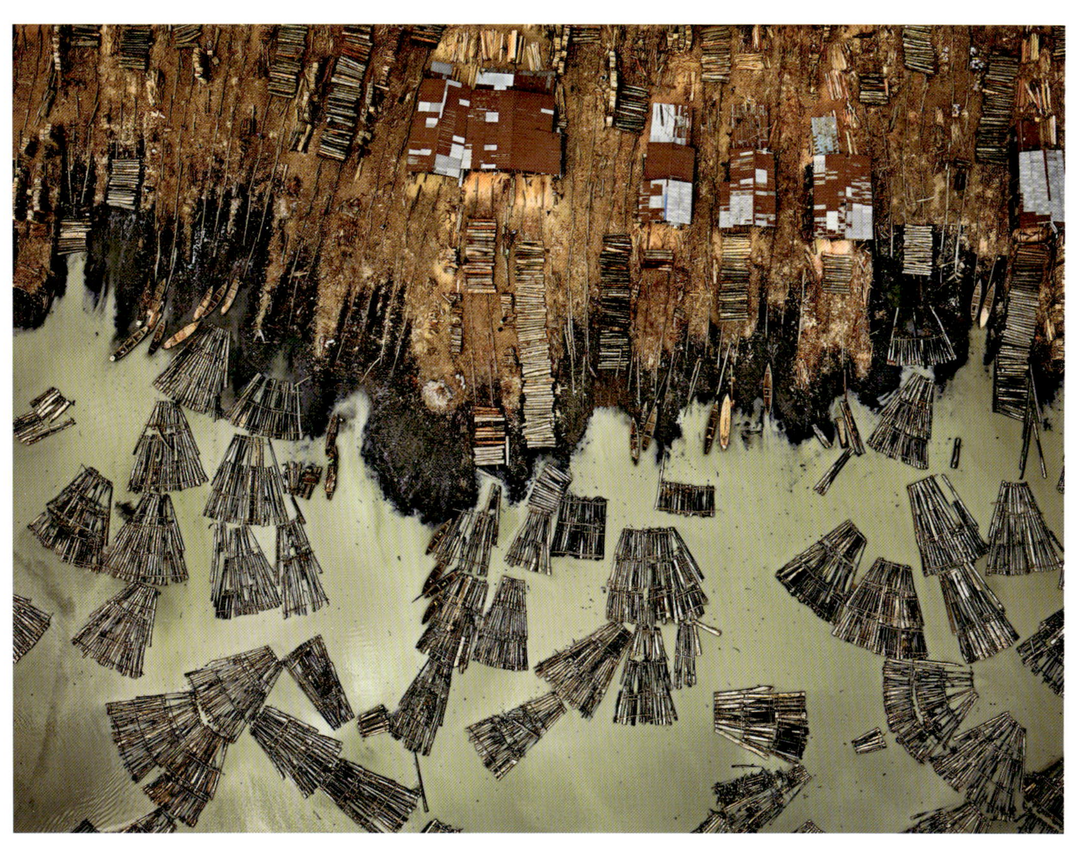

Edward Burtynsky, *Saw Mills #1, Lagos, Nigeria*, 2016

The otherworldly landscape of the Atacama is the driest non-polar desert on the planet.[1] Its picturesque salt flat rips tires to shreds and makes foot crossings practically impossible. Located amid this barren, surreal landscape is one of the largest known lithium reserves on the planet,[2] containing 27 percent of the world's lithium reserve base.[3] Receiving almost no rain, the Salar de Atacama has absorbed water from well outside its drainage basin for millions of years.[4] This water sank into the arid flats, creating ancient reserves of mineral-rich brine deep below the surface.[5] Declared a strategic resource by the Chilean government in the 1970s,[6] lithium is the core component of lithium-ion batteries. To extract lithium salts, brine is pumped up from the salt basin beneath the Salar.[7] It then evaporates in a series of artificial evaporation ponds, much like the artisanal systems seen in Gujarat, India (pages 110–111), or in the Bay of Cádiz in Spain (page 85).[8] As the liquid evaporates, the ponds move through a series of colour stages until their lithium concentration is determined to be high enough to be shipped for refining. Since the emergence of electric vehicle manufacturer Tesla, concerns have been raised that there will be a deficit in the lithium supply.[9] While this may bring about short-term price problems, lithium power has the potential to significantly reduce our carbon footprint.[10] It is now estimated that, by 2021, global lithium battery–making capacity will double, reaching 278 gigawatt hours of power annually.[11] With major powers like the United States and China actively working to ensure the development and expansion of the lithium supply, the metal is poised to become one of the most valuable commodities of the twenty-first century.

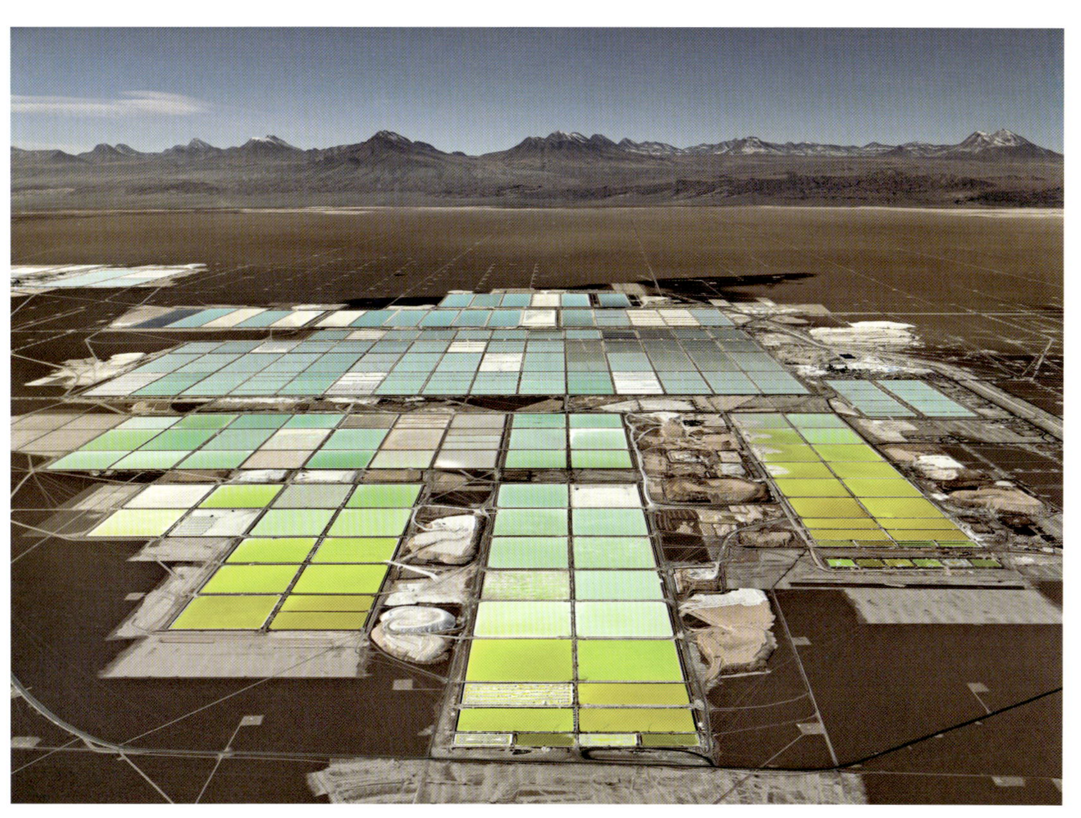

Edward Burtynsky, *Lithium Mines #1, Salt Flats, Atacama Desert, Chile,* 2017

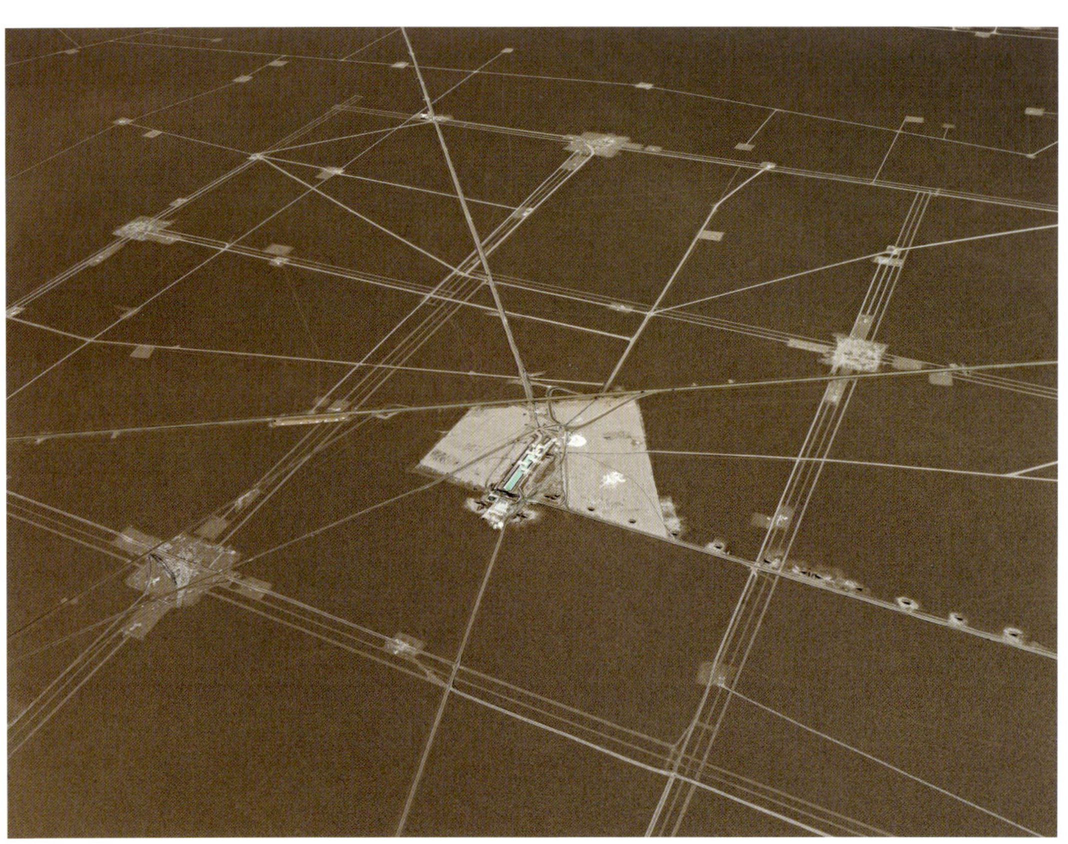

 Edward Burtynsky, *Brine Wells #1, Salt Flats, Atacama Desert, Chile*, 2017

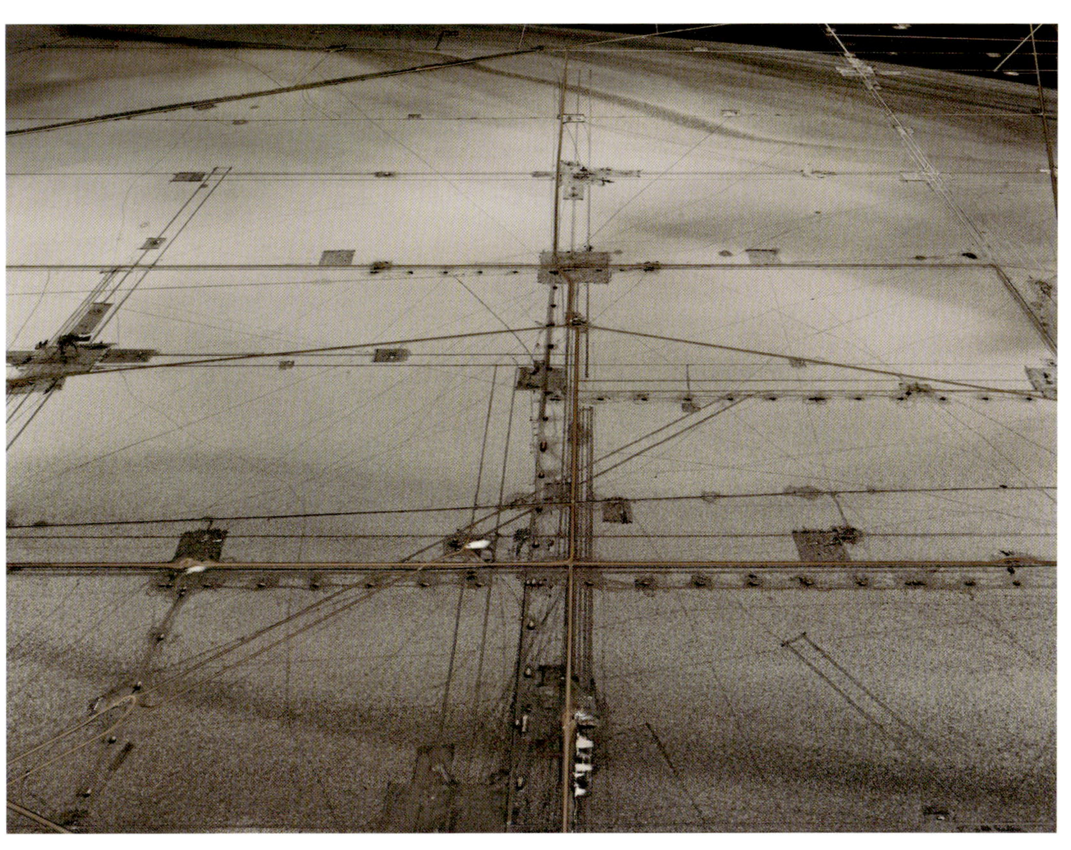

Edward Burtynsky, *Brine Wells #2, Salt Flats, Atacama Desert, Chile,* 2017

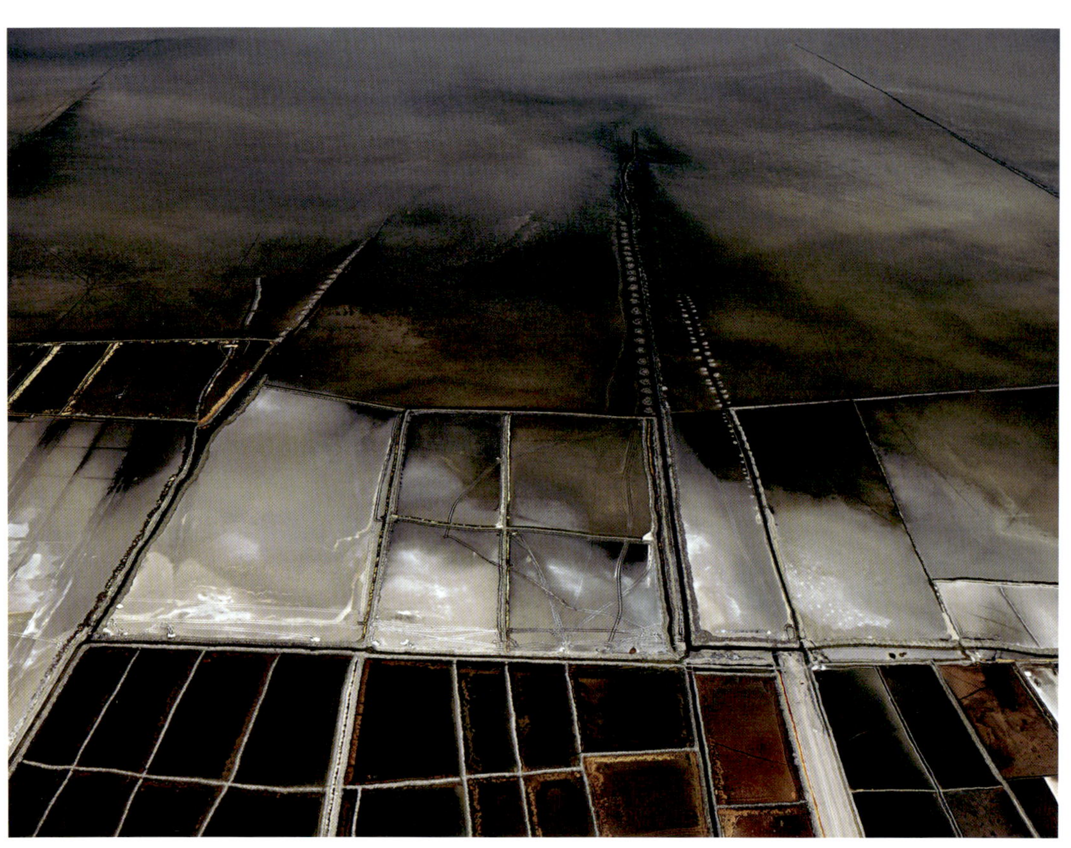

 Edward Burtynsky, *Salt Pan #21, Little Rann of Kutch, Gujarat, India*, 2016

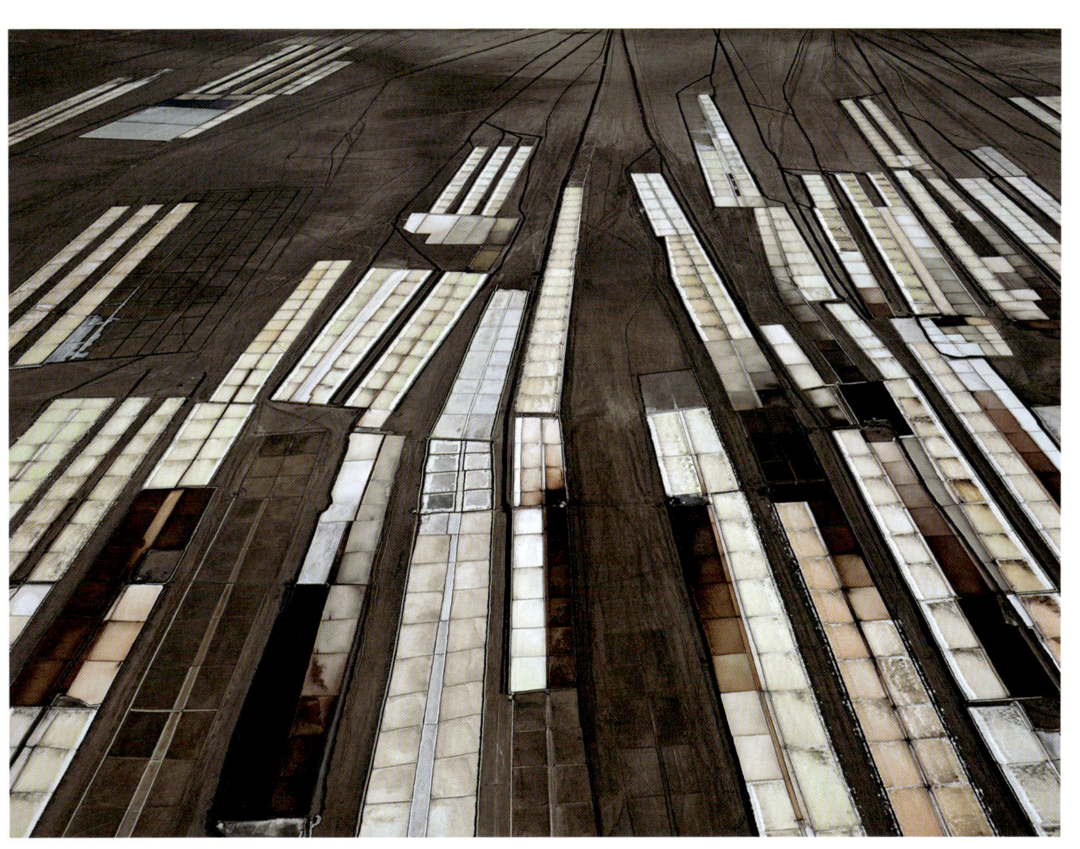

Edward Burtynsky, *Salt Pan #18, Little Rann of Kutch, Gujarat, India,* 2016

Seventy percent of the Nigerian government's revenue is tied up in the oil resources of the Niger Delta, with petroleum products making up 90 percent of all export revenues for the country.[1] Since oil was discovered in the region in 1956, the resource has proven to be both a blessing and a curse.[2] For decades, multinationals have abdicated responsibility for devastating oil spills onto land and into the water. Though Nigeria gained independence from British rule in October of 1960, much of the country's oil wealth continues to be diverted outside its borders. As a result, poor communities have begun pirating crude oil from pipelines—often with the support of local elites and politicians who reap benefits—through a process known as "bunkering."[3] Tapping into pipelines, makeshift micro-refineries are set up on a temporary or semi-permanent basis to turn the crude into low-grade fuel. Some government-commissioned estimates suggest that 250,000 barrels are stolen every day, but accurate figures are unknown.[4] Highly dangerous, these systems (and the broken pipelines that feed them) leak volumes of crude oil and toxic by-products into the surrounding forests and waterways. In order to build and access the pipelines, large areas of ancient forest are opened up through logging and burning. This, in turn, opens more rainforest and woodlands to (legal and illegal) logging.[5] Timber is then transported downriver using booms similar to those seen in British Columbia (page 97). For many of the logs, the final destination is Makoko, an informal community connected to Lagos (pages 72–73, 104–105). Here, sawmills process the hardwood for local use. Over time, biodiversity in the region has diminished, a result of continued deforestation and degradation.[6]
In 2010, Nigeria began the process of what *Foreign Affairs* described as "possibly the largest ever transfer of energy assets from foreign companies to local ones."[7] While the development of a localized industry certainly poses its own set of challenges, there is cautious hope that after years of corruption and violence through the collusion of government dictatorships and foreign corporations, Nigerians will start to see real benefits from the wealth of natural resources beneath their feet.[8] Still, bunkering by-products continue to pollute the ruined landscapes of former forests, offering long-term reminders of the impact of global consumer pressures on this land and its people.

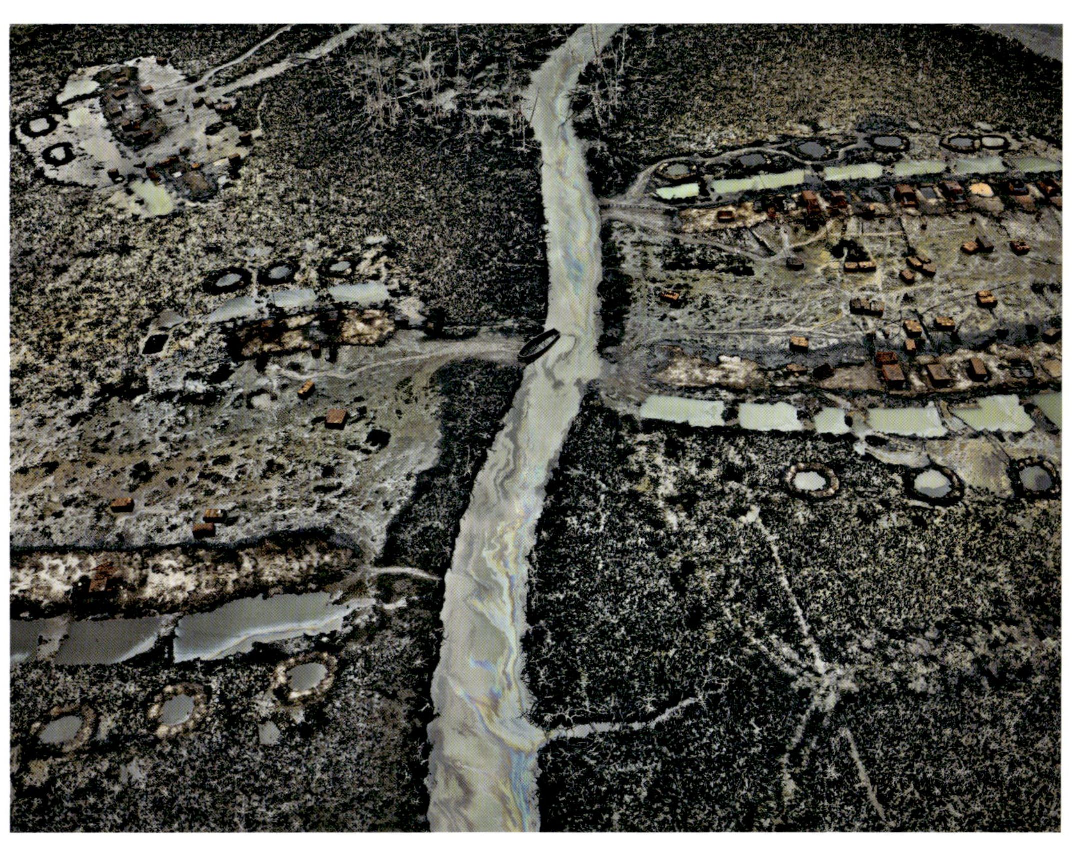

Edward Burtynsky, *Oil Bunkering #1, Niger Delta, Nigeria*, 2016

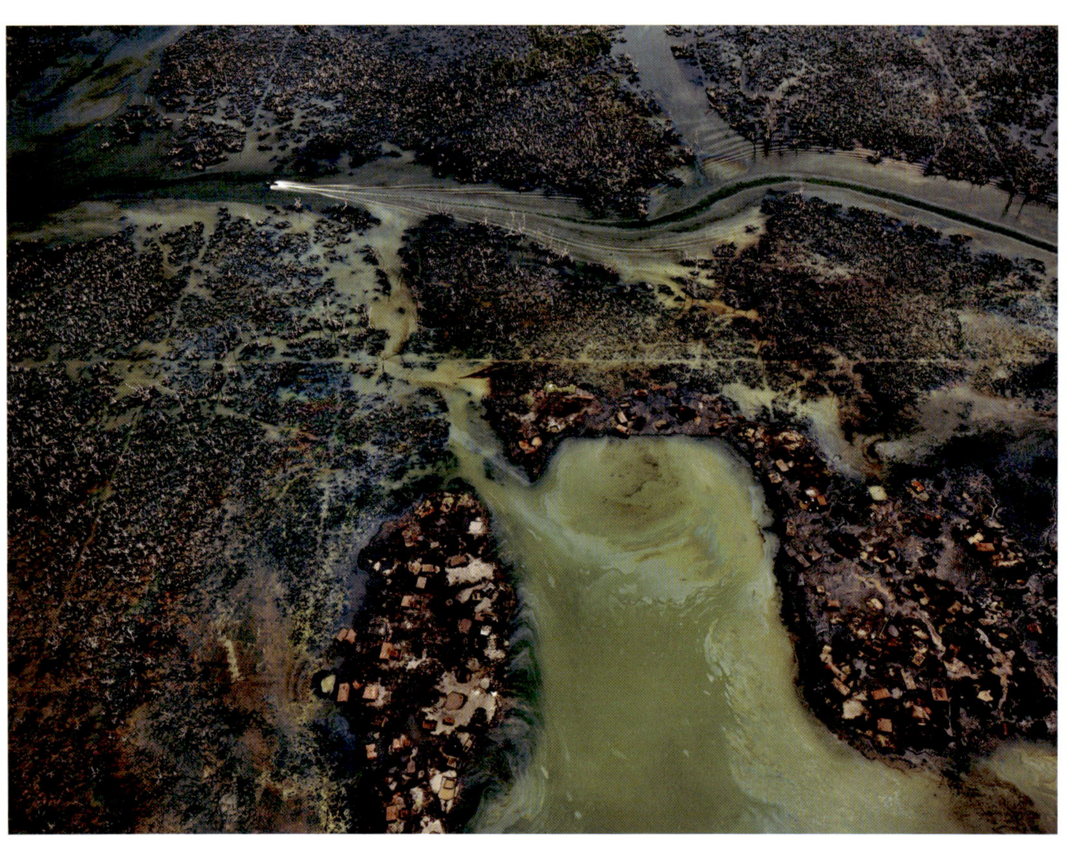

 Edward Burtynsky, *Oil Bunkering #4, Niger Delta, Nigeria,* 2016

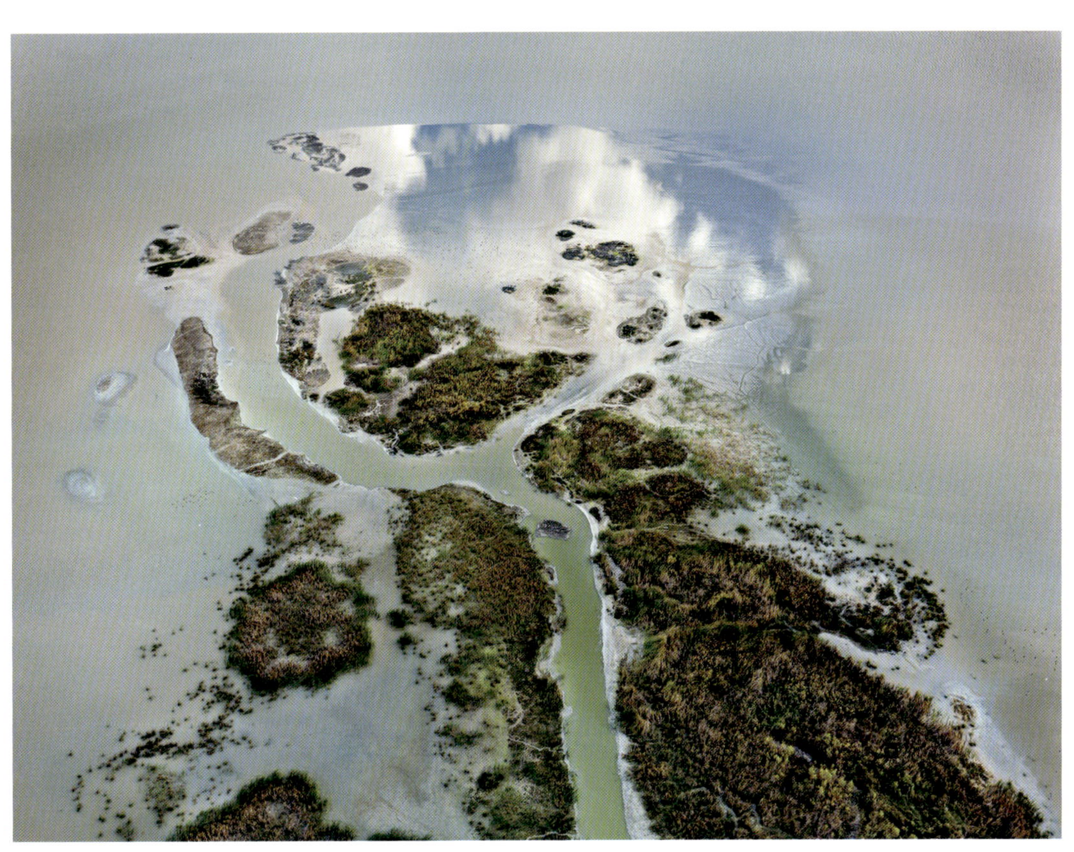

Edward Burtynsky, *Phosphor Tailings #6, Near Lakeland, Florida, USA*, 2012

Phosphate ore is one of the world's most critical mineral resources. Like fossil fuels, it is non-renewable. Phosphate rock (or phosphorite) deposits can be either sedimentary or igneous. The deposits mined in Florida are sedimentary, formed through the deposition of phosphate-rich materials in ancient marine environments.[1] To access the ore, phosphate surface-mining operations must clear natural vegetation and topsoil.[2] These regions are typically unable to revert back to their natural state following this polluting process, which involves the excavation and deposition of overburden and related waste disposal, thus leaving a nutrient-deficient landscape in its wake.[3] Phosphates, essential to all life, are crucial to industrial agriculture. A key component in DNA, phosphorus is necessary for plant energy metabolism and growth.[4] However, water systems large and small are threatened by its nutrient-rich agricultural runoff. While phosphate rock is insoluble in its untreated form, when treated with sulphuric acid, it produces phosphoric acid—the water-soluble substance from which most phosphate fertilizers are derived.[5] Runoff, from both the industrial mining of phosphates and their use in agriculture operations, is a key factor in the creation of harmful algal blooms, loss of fish species, and the pollution of vital water sources. Looking back in geological time, high levels of phosphorus have been associated with oceanic anoxic events (OAEs). Such events occur at times with low levels of dissolved oxygen in the ocean, and tend to coincide with both "greenhouse" climates and mass extinctions, like the Permian–Triassic extinction event (nicknamed "The Great Dying"). Shortly before the advent of wide-scale use of fertilizers, Sir George Knibbs's 1928 book, *The Shadow of the World's Future,* suggested that beyond a population of 7.8 billion, agricultural land would have to be made significantly more productive.[6] Soils are a kind of contained energy, with all the nutrients necessary to life contained within them. Once those nutrients are gone, soil has no capacity for growth. There is no substitute for phosphates in industrial agriculture. With an expanding world population, demand for this essential mineral will only increase, putting ever greater pressure on already stressed ecosystems.

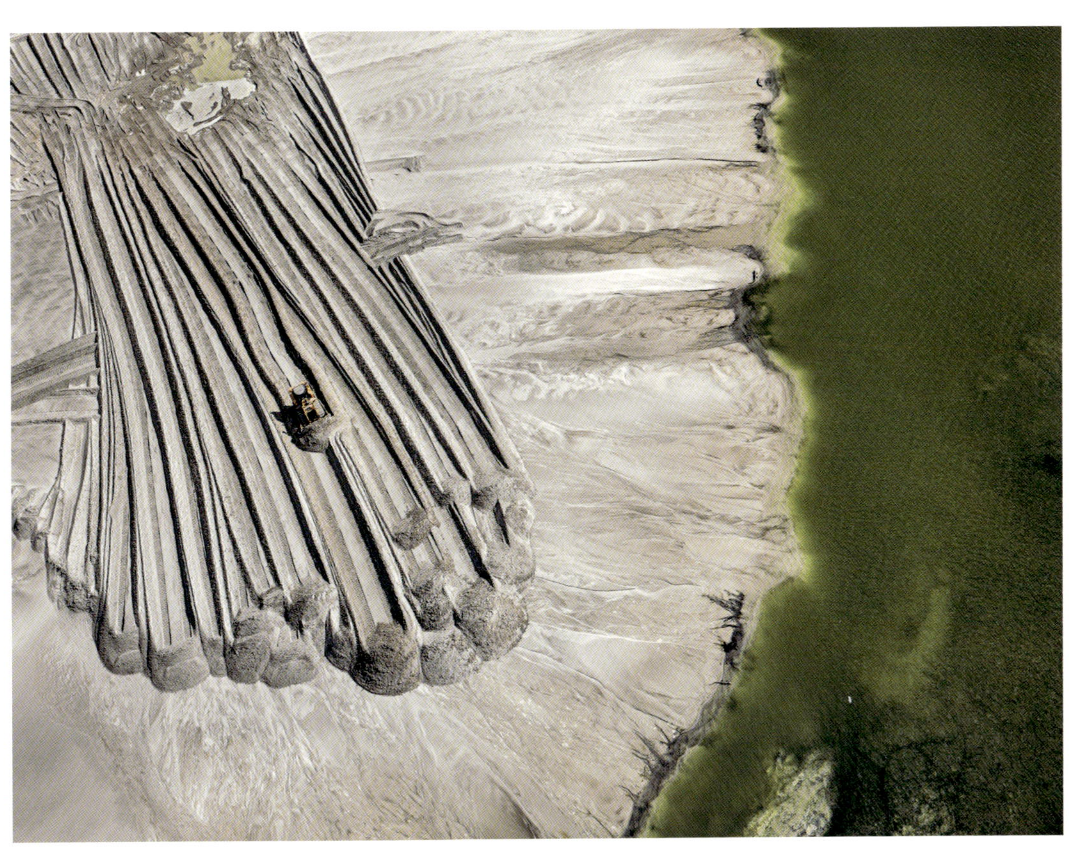

Edward Burtynsky, *Phosphor Tailings Pond #4, Near Lakeland, Florida, USA*, 2012

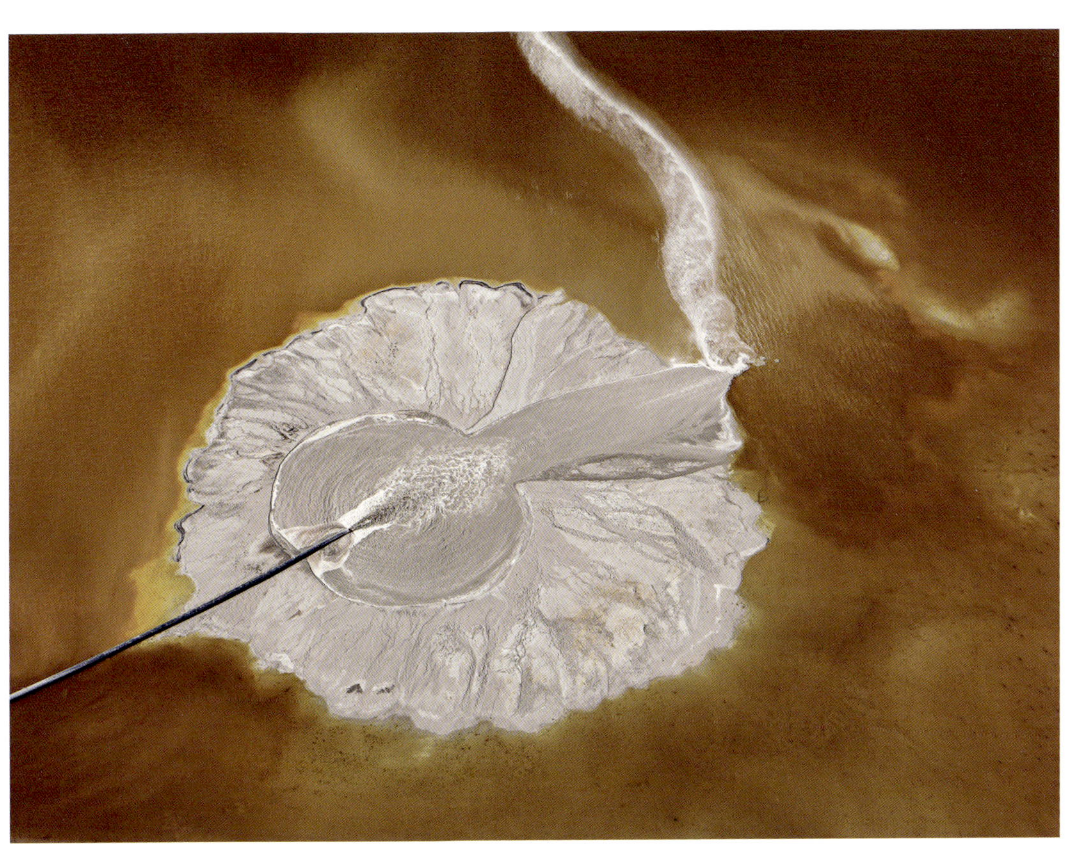

Edward Burtynsky, *Phosphor Tailings #5, Near Lakeland, Florida, USA,* 2012

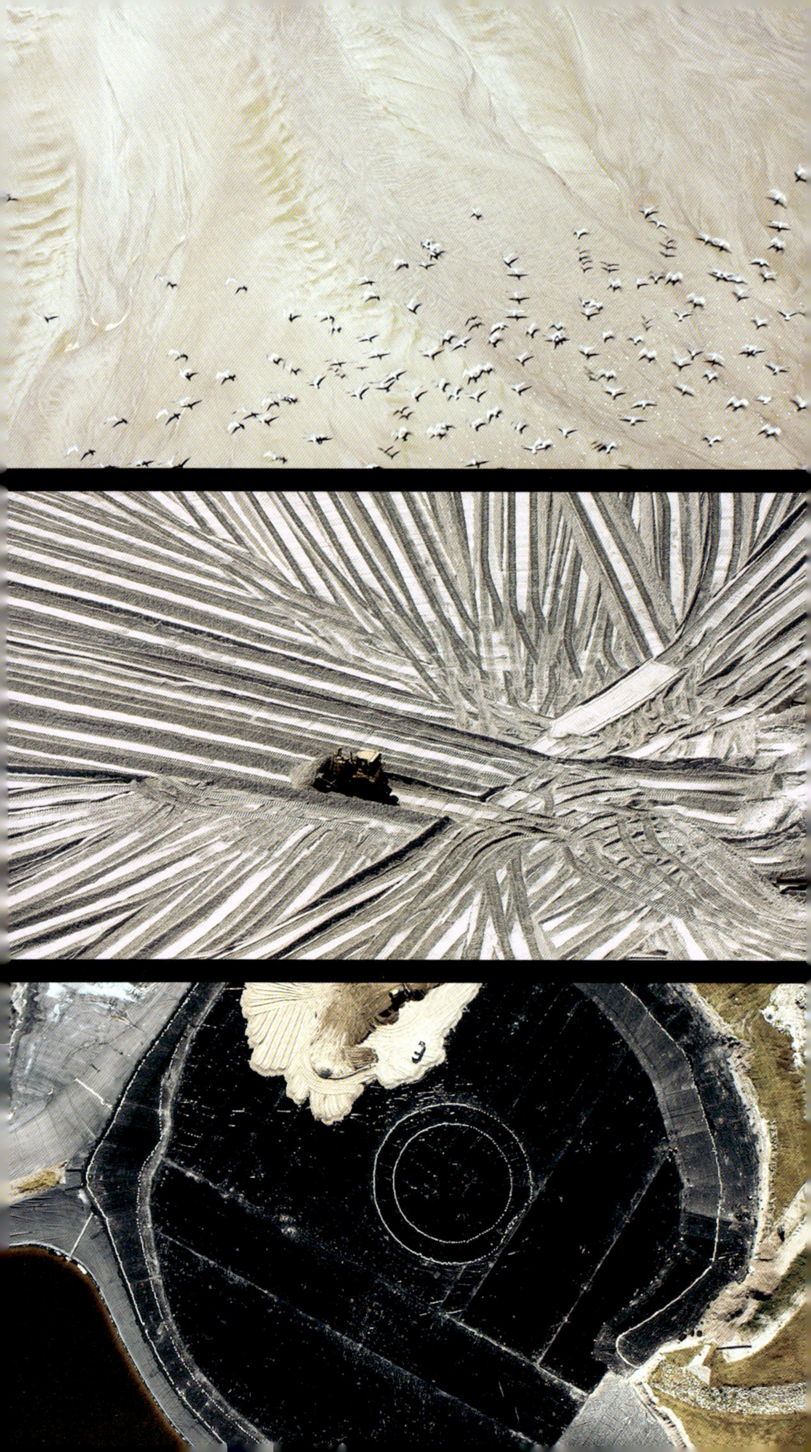

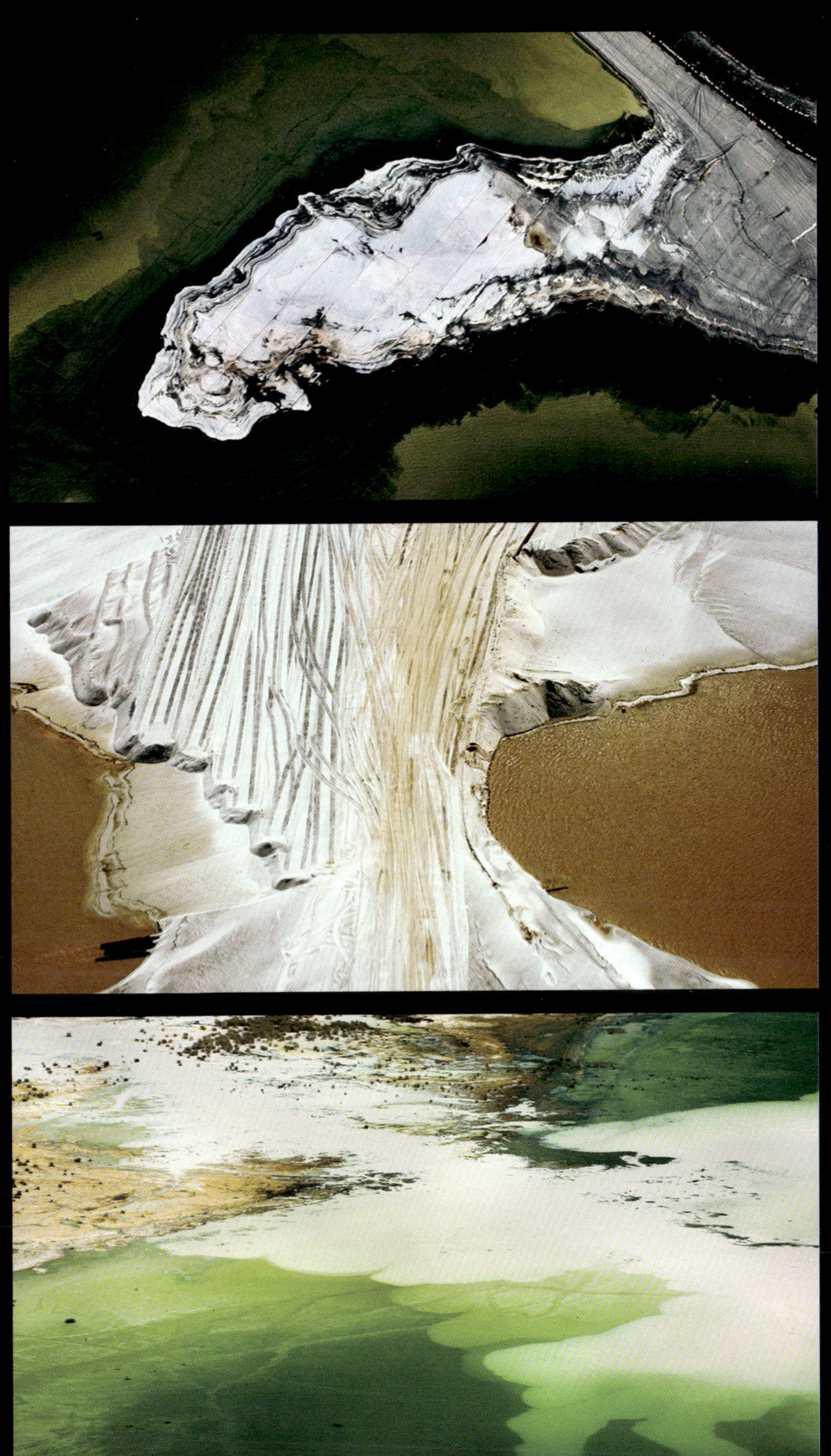

Jennifer Baichwal and Nicholas de Pencier, *Phosphate Mines, Florida, USA* (film stills), 2018

The Hambach mine is the largest open-pit mine in Germany.[1] The coal at this mine can only be reached by removing mountains of unwanted material, which is then dumped on the land to form a massive man-made hill, a task that requires unique heavy machinery. The mine at Hambach features the Bagger 291 and 293, which are among the largest land vehicles in human history by weight.[2] Bucket-wheel excavators such as these remove the sandy overburden covering the coal. At a rate of 240,000 cubic metres of material per day, the Bagger 293 removes approximately 220 to 250 million cubic metres of overburden every year.[3] The entire machine stands at 220 metres in length and 94.4 metres in height.[4] It has eighteen buckets in total, each of which can hold more than 5 cubic metres of sand or coal. The lignite, or soft brown coal, produced at the 85-square-kilometre Hambach mine accounts for approximately 5 percent of the country's energy.[5] Lignite is a particularly inefficient and dirty type of fuel. As incidents of protest increase,[6] the future of extraction in the region is unclear. Despite significant investments in renewable energy, Germany still leads the EU for carbon emissions as of 2017.[7]

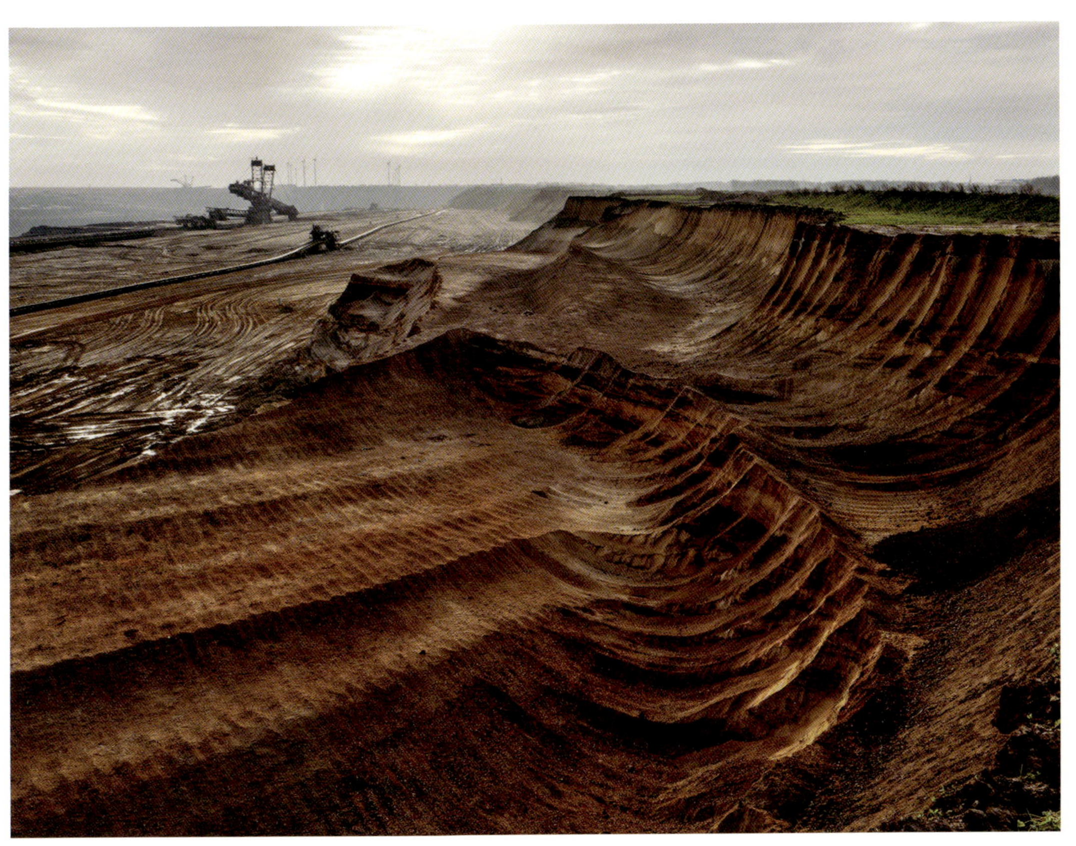

Edward Burtynsky, *Coal Mine #1, North Rhine, Westphalia, Germany,* 2015

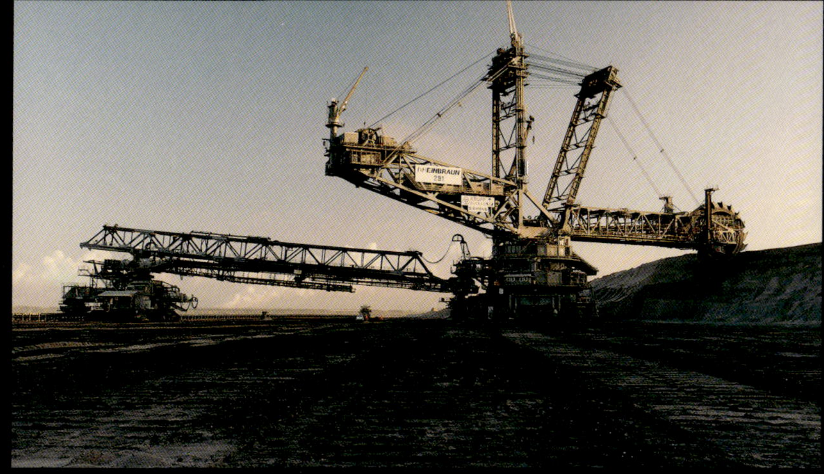

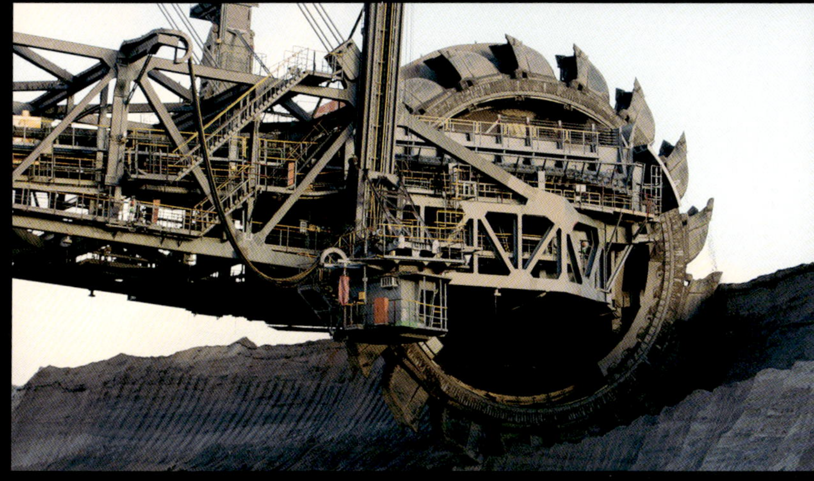

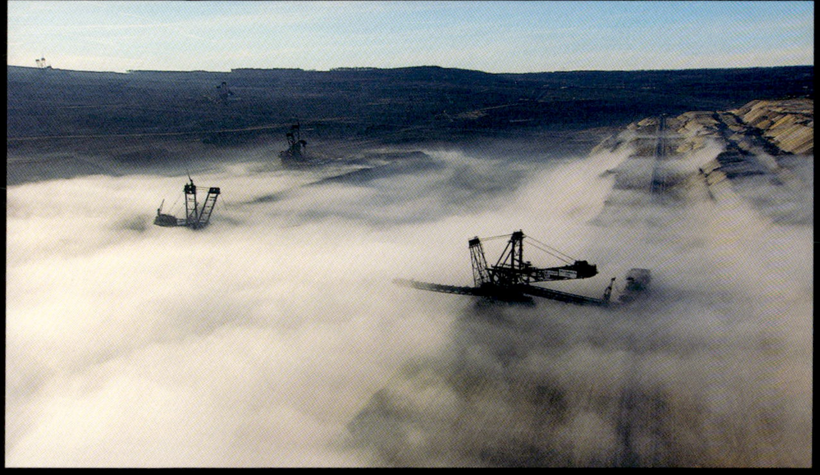

Jennifer Baichwal and Nicholas de Pencier, *Bagger 291, Hambach Lignite Mine, Germany* (film stills), 2018

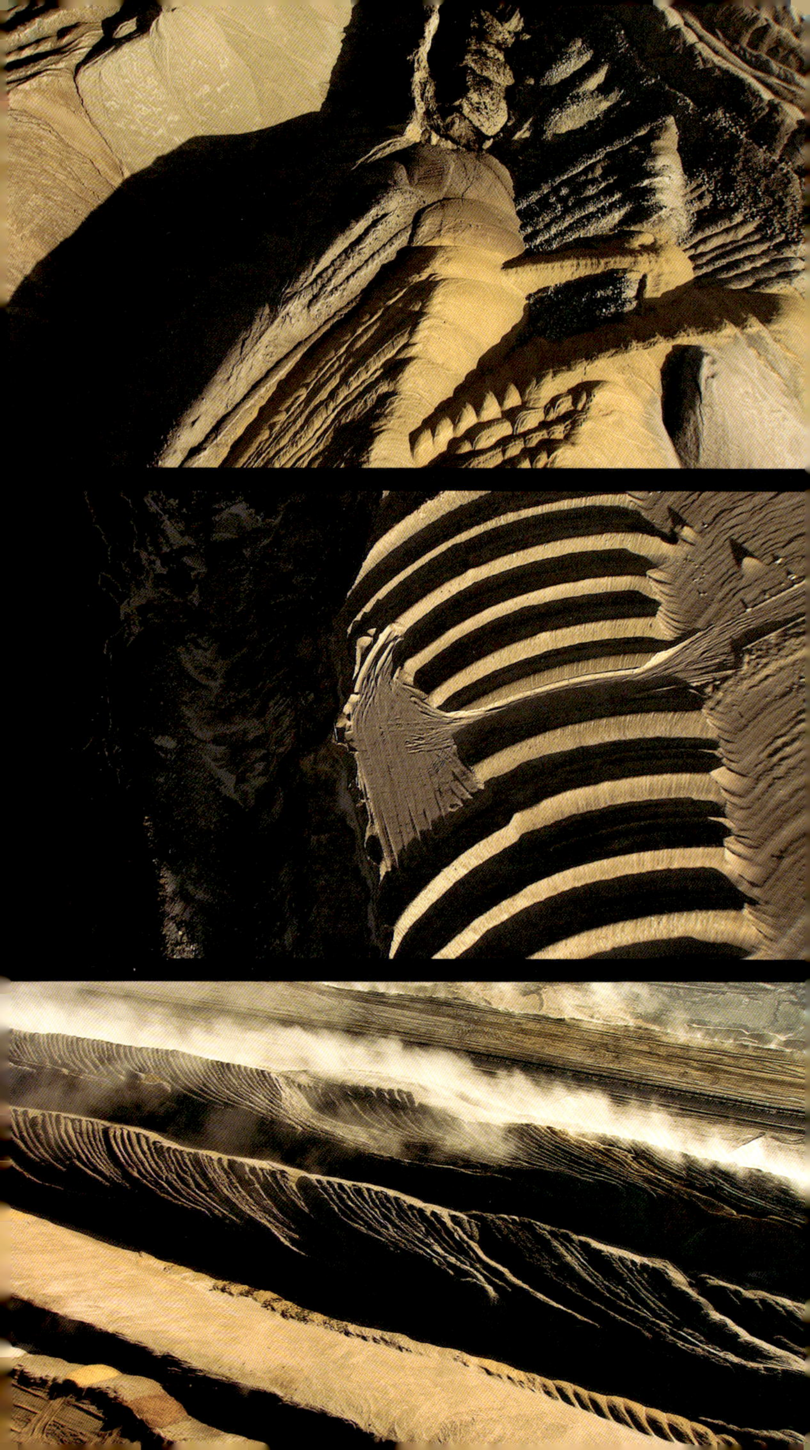

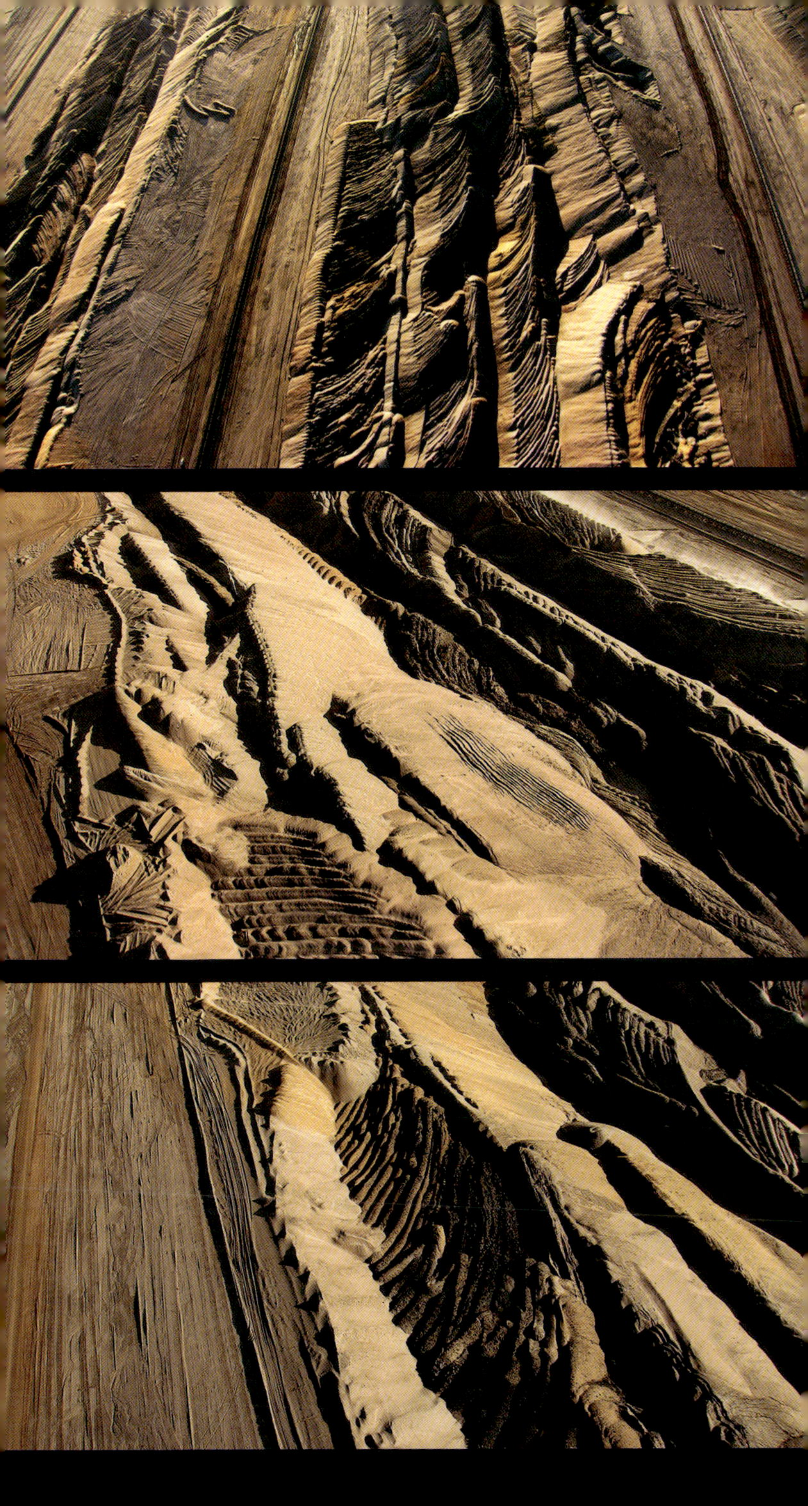

Jennifer Baichwal and Nicholas de Pencier, *Hambach Lignite Mine, Germany* (film stills), 2018

These images depict coal mining and natural gas fracking between Casper and Gillette, Wyoming. In 2009, the United States surpassed Russia as the world's largest producer of natural gas, and in 2013 it surpassed Saudi Arabia to become the world's top producer of petroleum hydrocarbons.[1] As of 2016, it remained the leader in both categories.[2] The country's dominance is largely due to the development of multi-stage hydraulic fracturing (or fracking), a process wherein water, sand, and chemicals are pumped deep into bore wells to crack open fissures and bolster the flow of oil and gas.[3] This method of oil production, which requires approximately 19 cubic metres of water per well,[4] has the potential to cause significant groundwater contamination, primarily through processes that have received minimal study or regulation. By 2019, these same technologies will make the U.S. the world's largest producer of petroleum, overtaking Saudi Arabia for the first time.[5] Since the Industrial Revolution, over 390 billion tonnes of anthropogenic carbon emissions have been released into the air through cement production and the burning of fossil fuels.[6] For much of the earth's history, CO_2 levels ranged between 200 parts per million (ppm) and 280 ppm. In 2013, CO_2 levels reached 400 ppm for the first time on record.[7] At our current rate, it is estimated that the earth's climate will warm from 3.2°C to 5.4°C above pre-industrial levels by 2100.[8] The discovery and exploitation of new methods of oil extraction ensure that until a more efficient form of energy is found, global sustainability will take a back seat to the status quo.

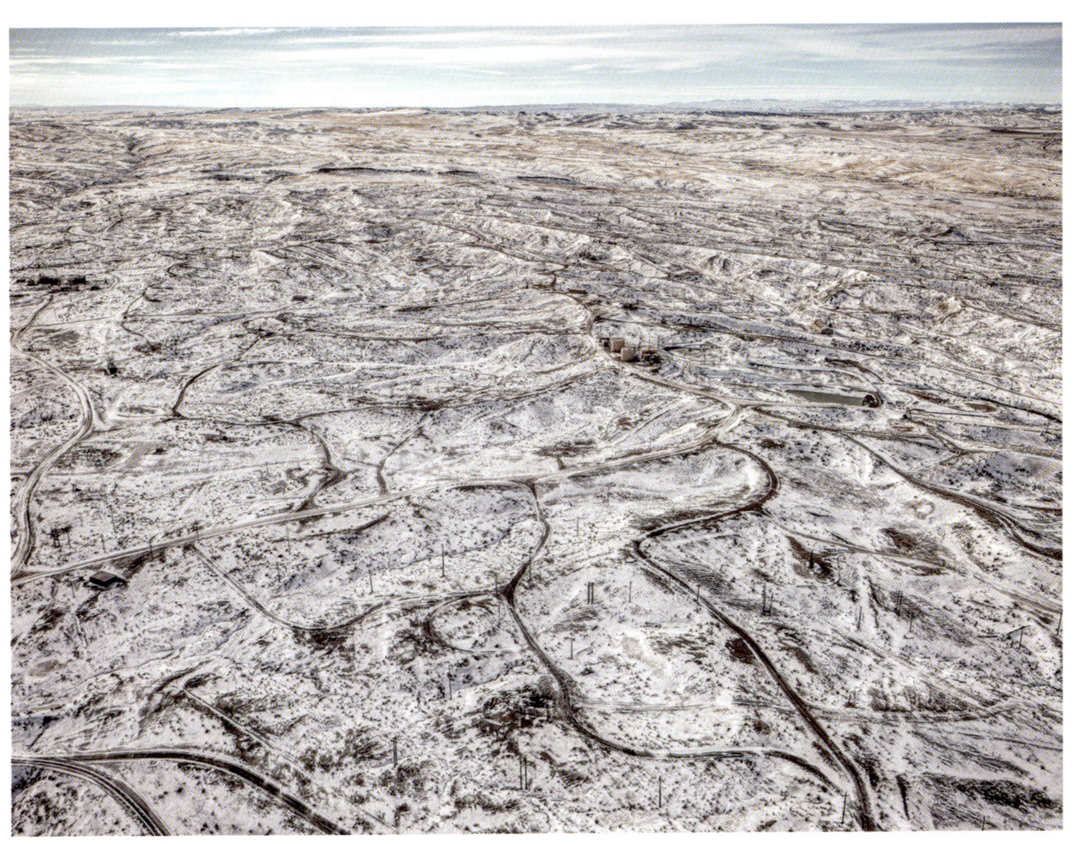

Edward Burtynsky, *Fracking, Near Gillette, Wyoming, USA*, 2015

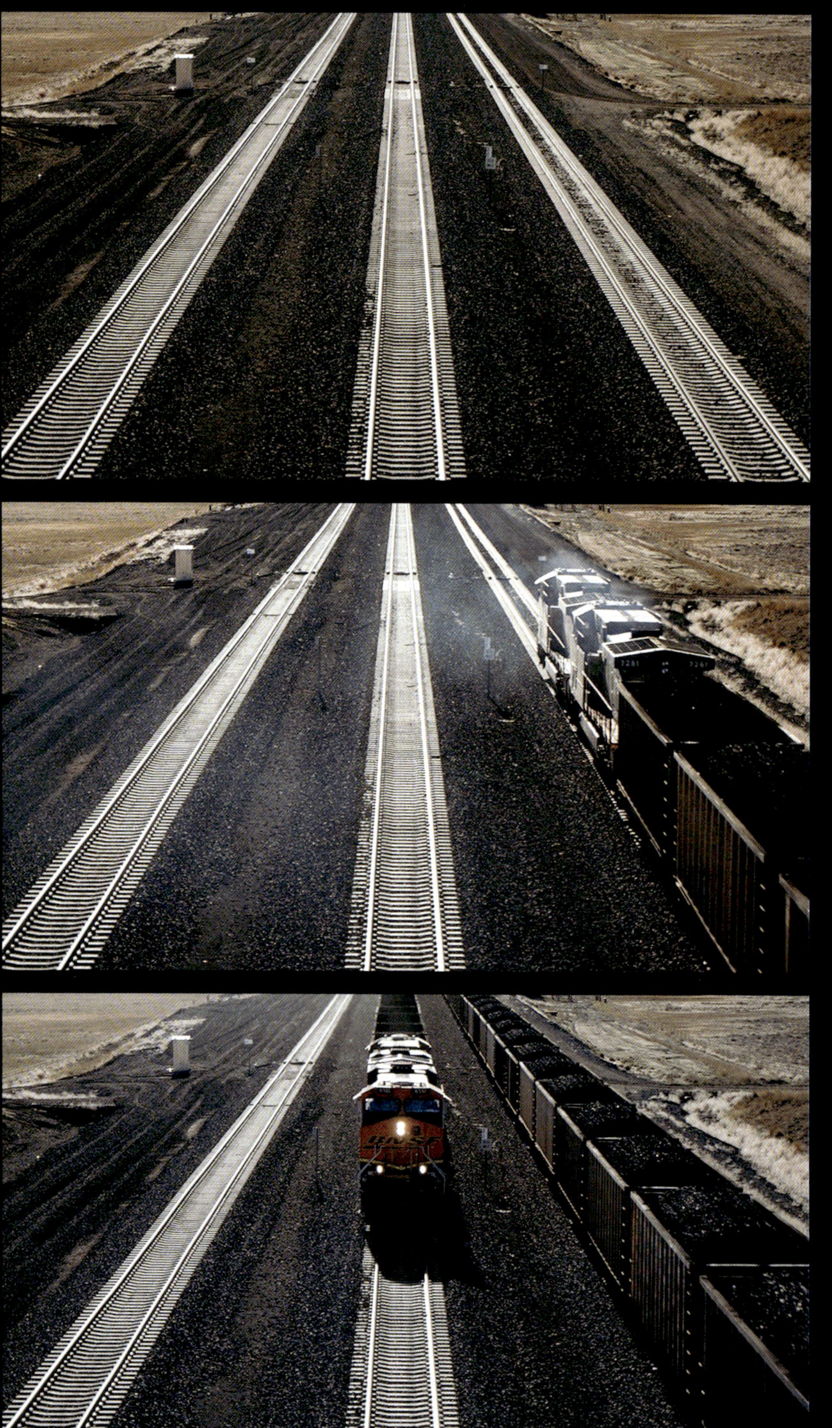

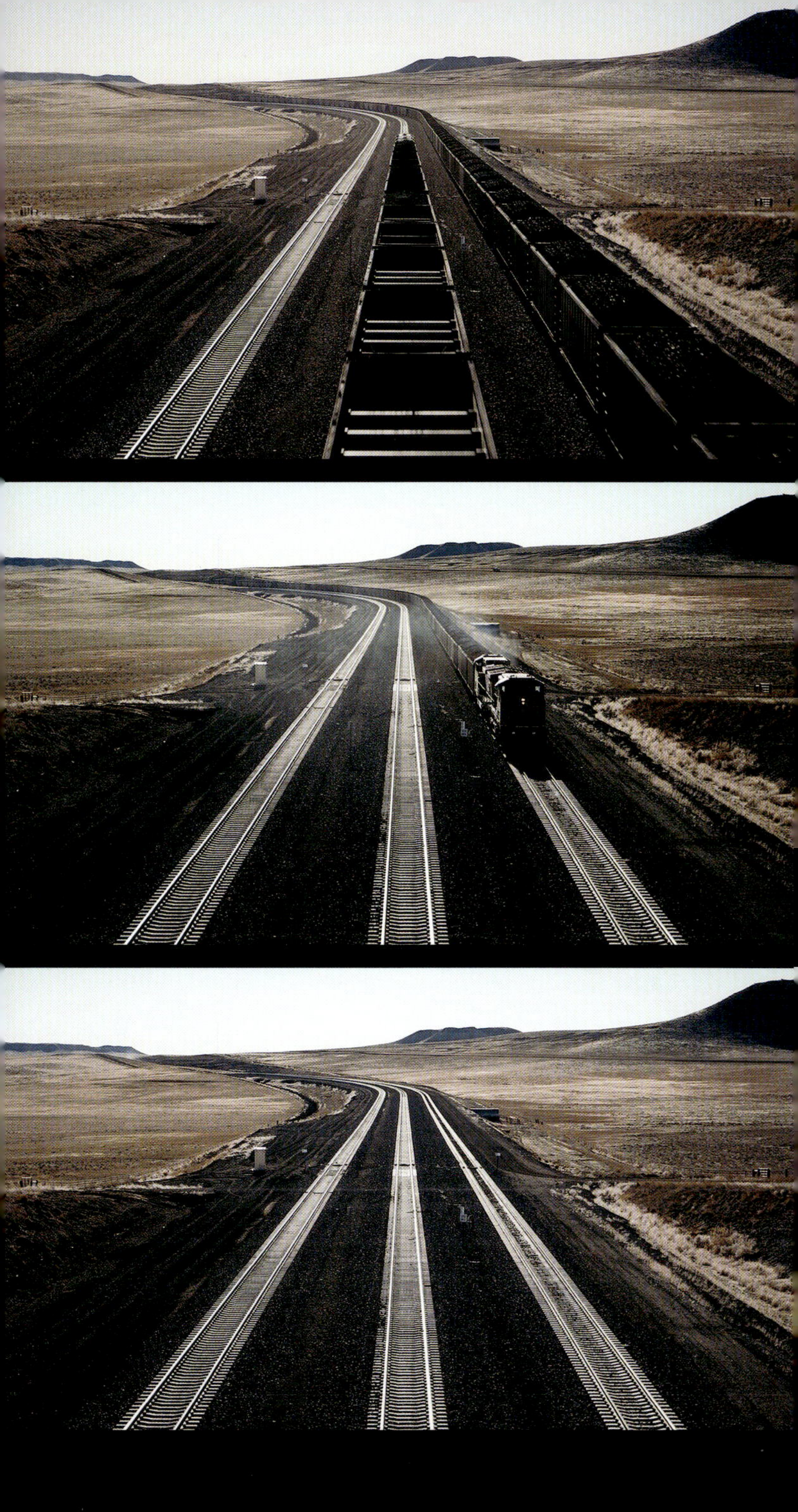

Jennifer Baichwal and Nicholas de Pencier, *Coal Trains, Wyoming, USA* (film stills), 2018

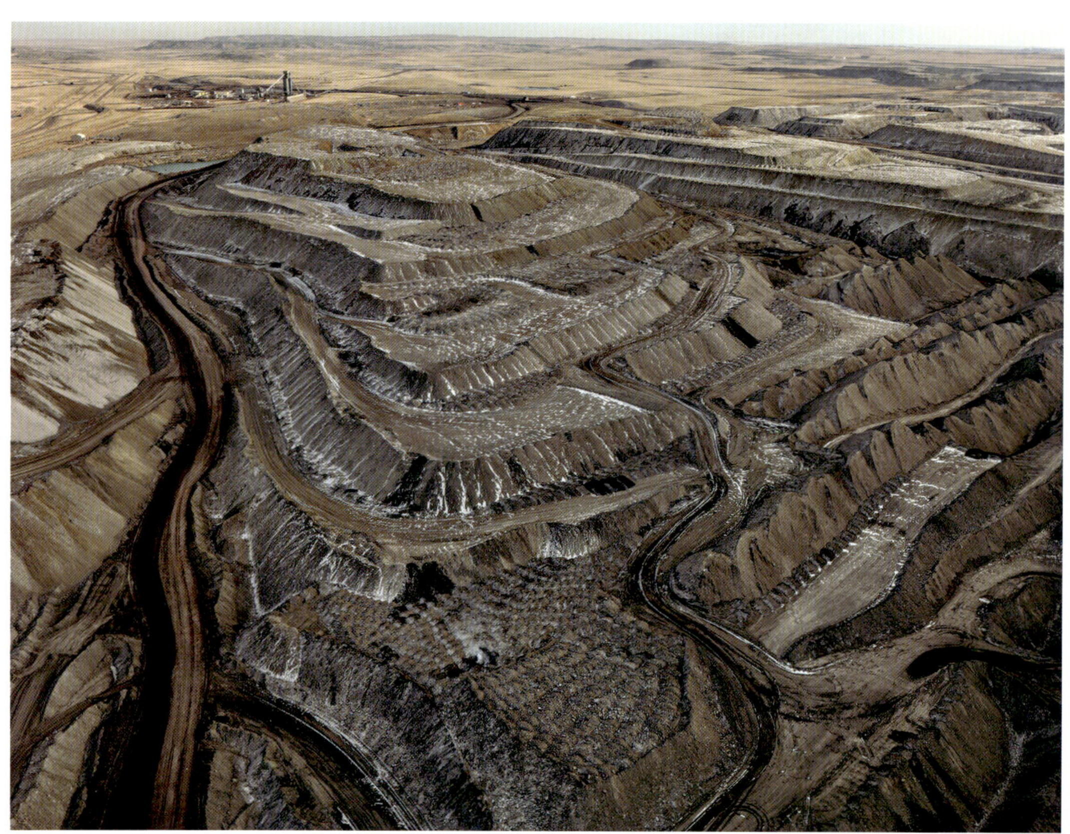

 Edward Burtynsky, *Coal Mining, Near Gillette, Wyoming, USA*, 2015

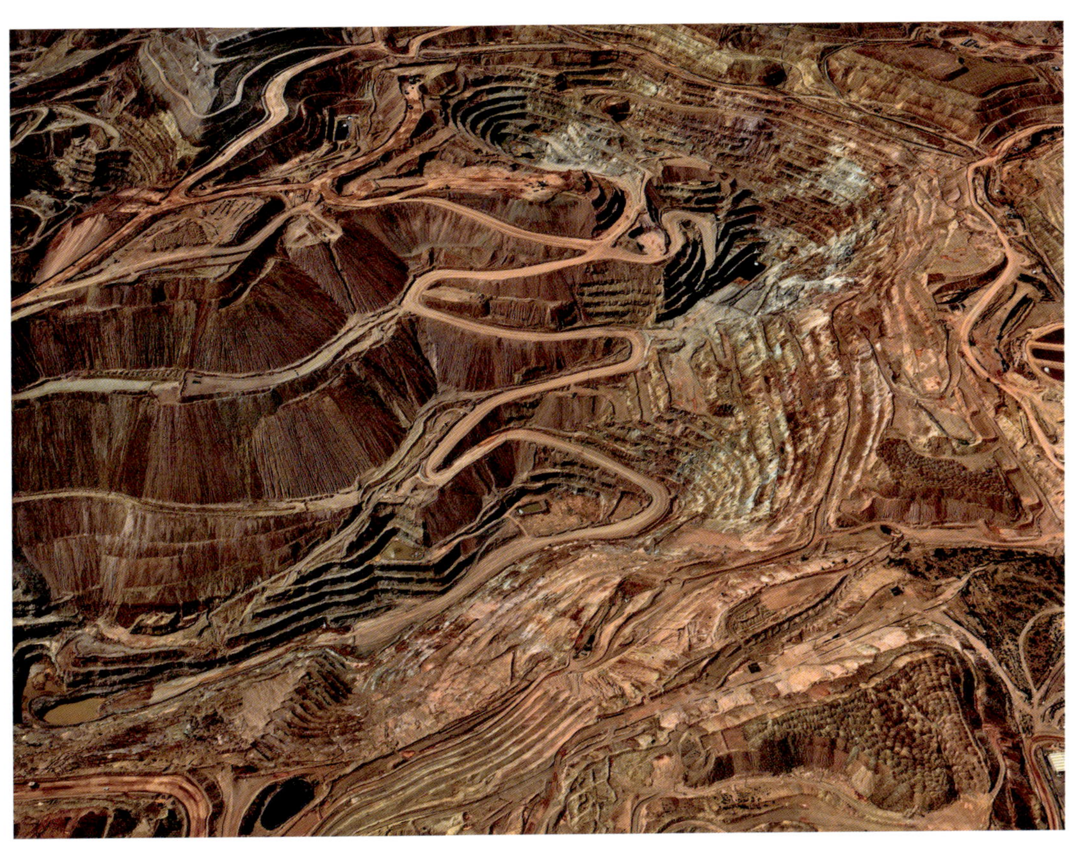

Edward Burtynsky, *Tyrone Mine #3, Silver City, New Mexico, USA*, 2012

The Chuquicamata copper mine, popularly known as Chuqui, has been in continuous production for over a century. It is one of the world's largest open-pit mines, with the capacity to process 350 kilotons of ore annually in its onsite smelters.[1] Copper has been mined by humans in this region for millennia[2] and Chile has been the world's largest supplier of copper since the year 2000.

Copper mining today, as in the past, is primarily done through open-pit methods. The difference now is the incredible scale. Chuquicamata is situated on 120 square kilometres of land,[3] the mine itself being 4.5 kilometres long and 3.5 kilometres wide, with an open-pit depth of almost a kilometre.[4] In 2012, after a century of industrial mining, construction began on an underground mine to extract Chuquicamata's remaining reserves. Projected figures from the site suggest that Chuqui will eventually produce an estimated 140,000 tonnes of ore per day, as well as 366,000 tonnes of copper and 18,000 tonnes of fine molybdenum per year.

In 2009, the United States Geological Survey estimated that every American born in 2008 will use 593.7 kilograms of copper throughout his or her lifetime.[5] As a highly recyclable material, recovered copper takes 85 to 90 percent less energy to produce than copper that is newly mined from the earth.[6] The value of new copper as opposed to its recycled counterpart, however, ensures that global demand will encourage mines to be dug until their resources are exhausted.

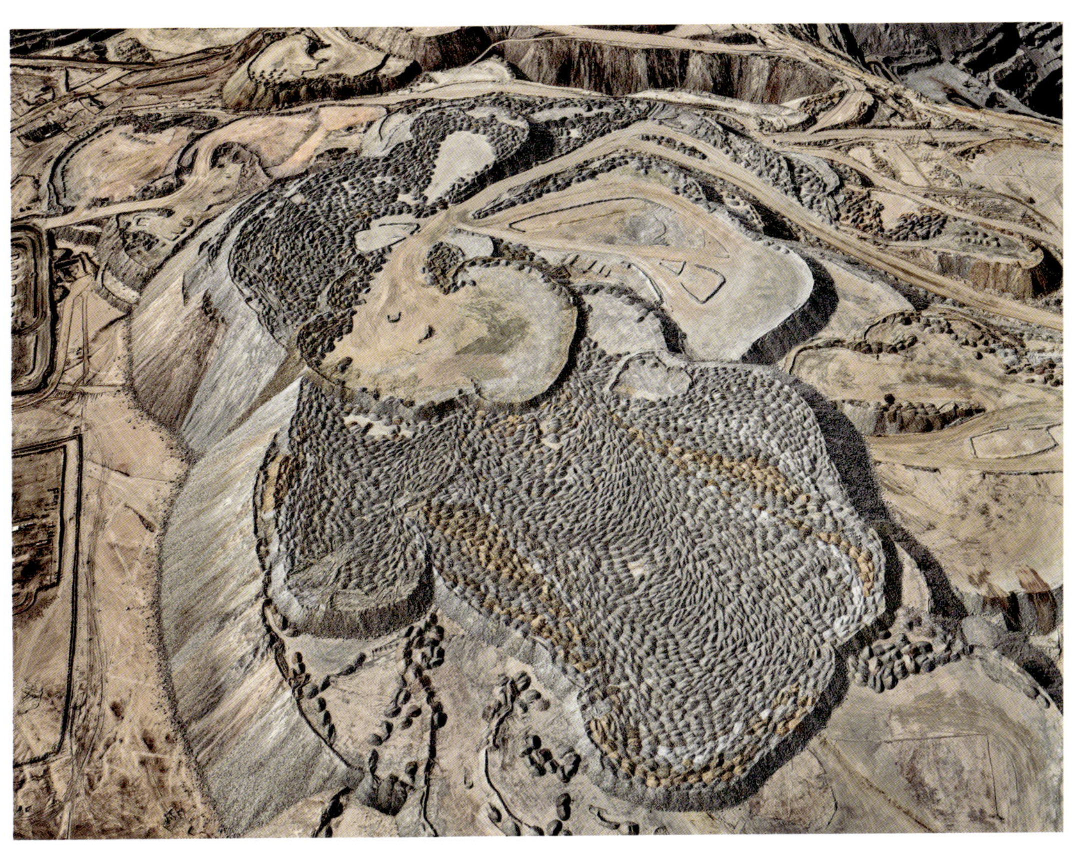

Edward Burtynsky, *Chuquicamata Copper Mine Overburden #1, Calama, Chile,* 2017

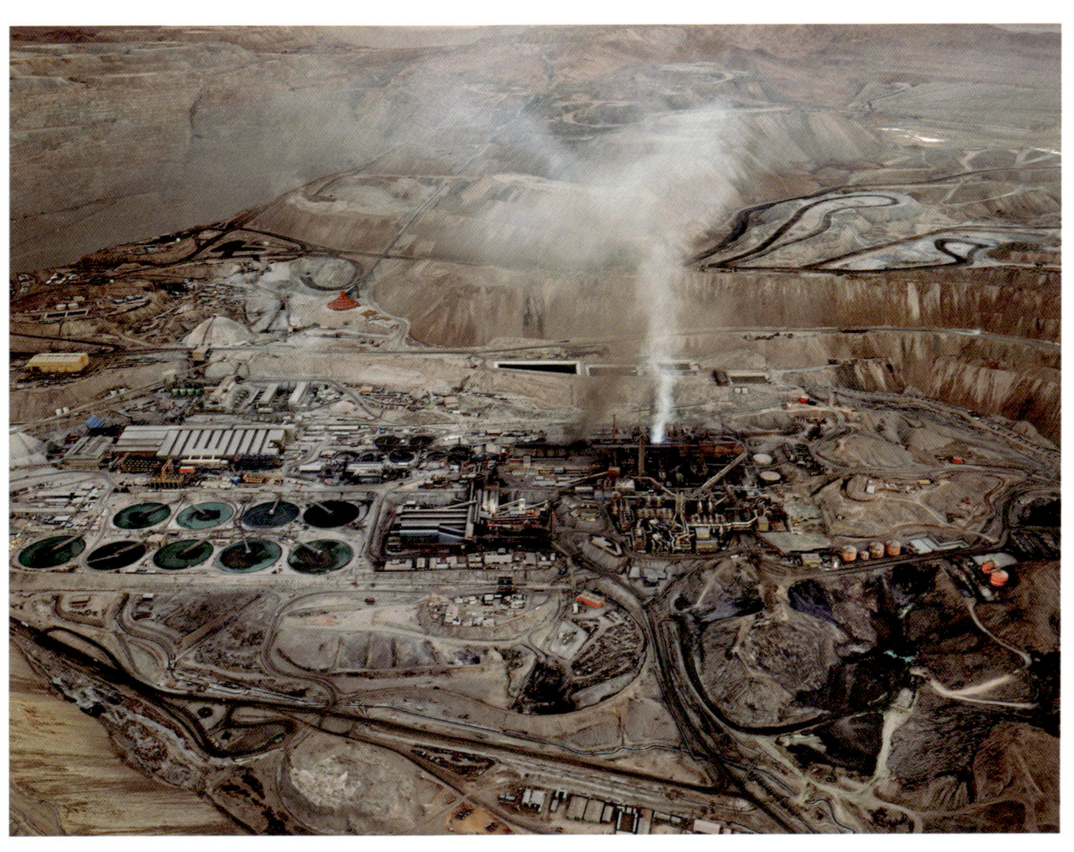

Edward Burtynsky, *Chuquicamata Copper Concentrator and Smelter, Calama, Chile*, 2017

Like most extractive practices, copper smelting is highly water-intensive, requiring between 1,500 and 3,000 litres of water for every ton of processed ore.[1] This can be seen in the aerial photographs from Arizona, the United States' primary copper producing region, in which large tailing ponds contain liquid reserves of the effluents left by the copper extraction process. The swirling, marble-like colours are the result of leached heavy metals.[2] Overburden (all soil above the bedrock) and waste rock (the portion below the cut-off grade for copper) are also removed in heap leaching, the method most commonly employed in the United States.[3] A significant amount of the water is recovered from tailings ponds and recirculated through the industrial process, but about 750 litres of fresh water is still required for every ton of ore.[4] At some mines, systemic improvements have reduced that amount to 340 litres per ton.[5] With most operations located in arid environments, every litre counts.

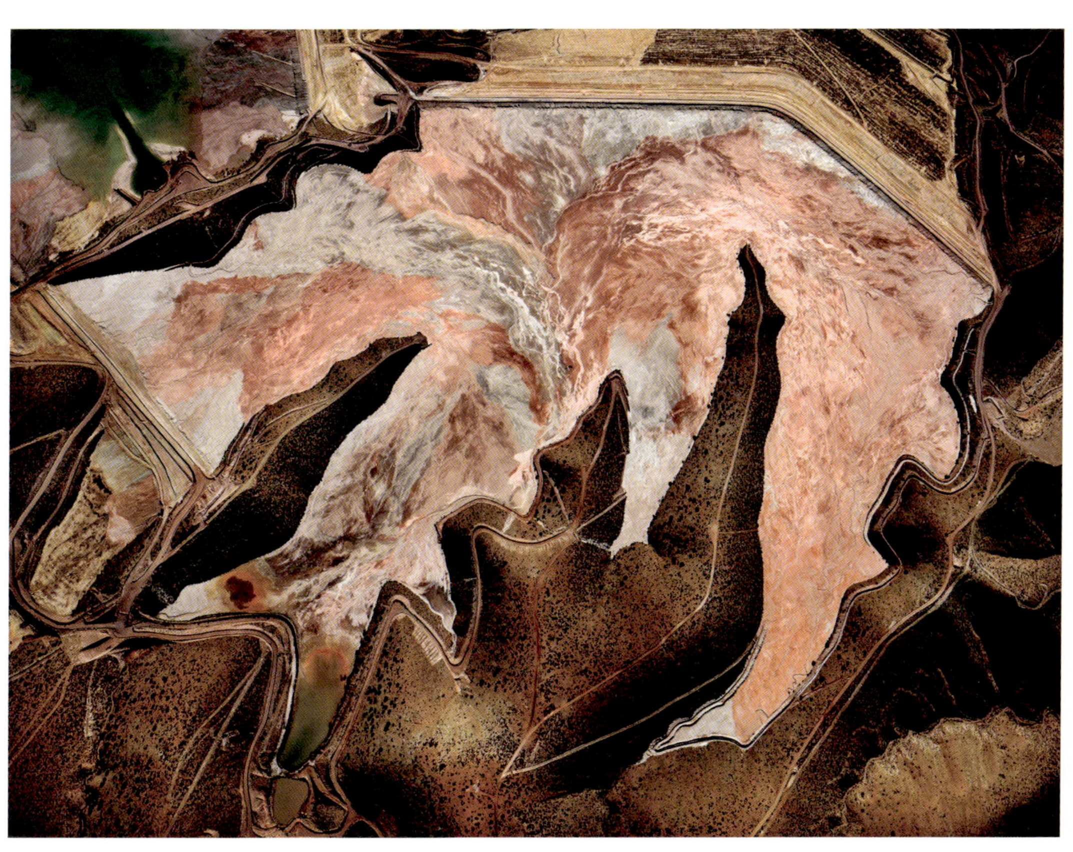

Edward Burtynsky, *Morenci Mine #1, Clifton, Arizona, USA*, 2012

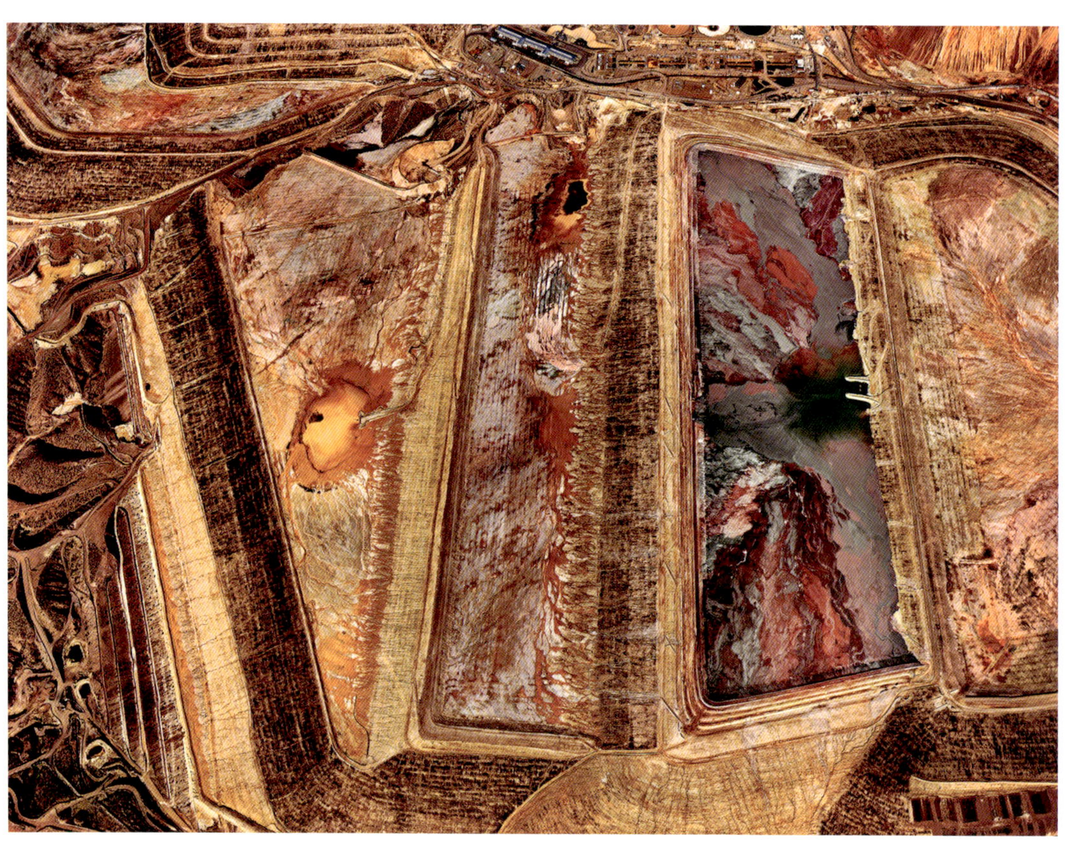

 Edward Burtynsky, *Morenci Mine #2, Clifton, Arizona, USA*, 2012

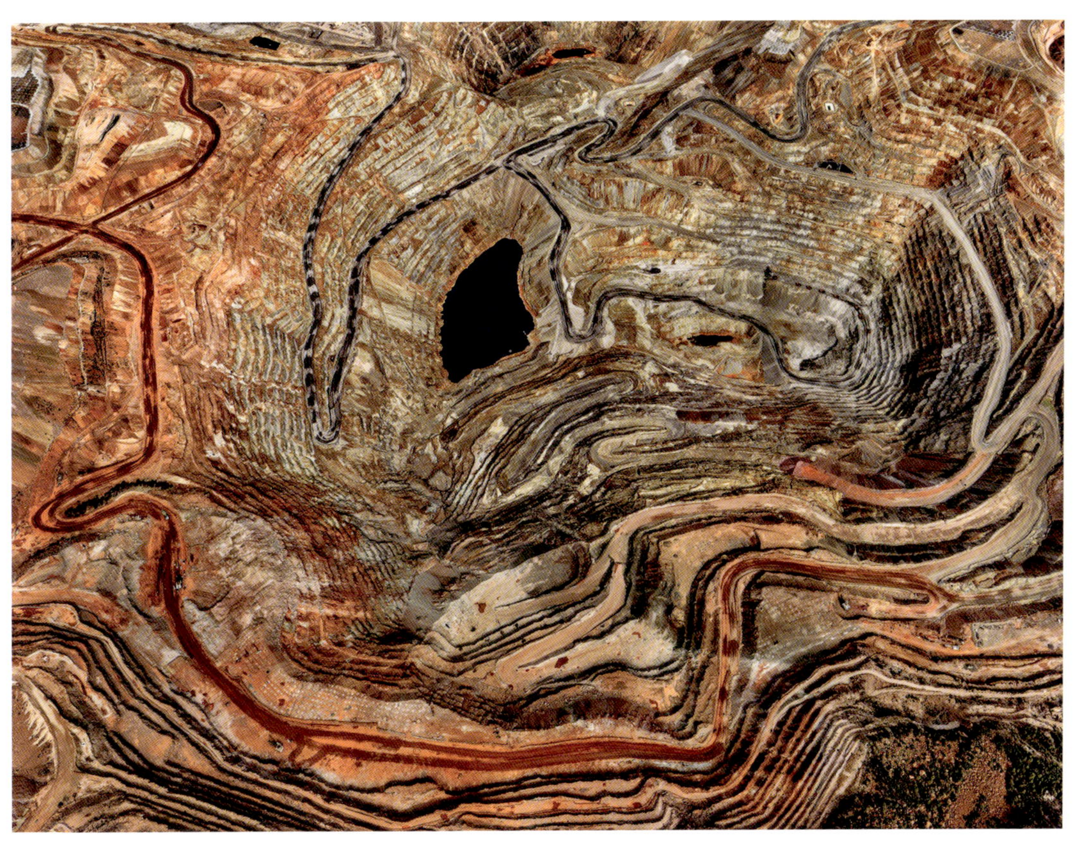

Edward Burtynsky, *Chino Mine #5, Silver City, New Mexico, USA*, 2012

Up to four hundred metres beneath Berezniki, Russia, tunnelling machines, referred to as "combines," reveal vividly coloured layers from an ancient sea floor. While the strata of Zumaia (pages 60–61) were raised above sea level, these mineral salts remained underground. Today, they have been revealed through the mining of potash—an indispensable fertilizer. As the combines pass through, they leave behind impressions in the soft rock that can look like fossils of the ancient sea life from which they were formed. These colourful walls contain the minerals that make up potash—a combination of halite, carnallite, and sylvite. Completely enveloped in darkness, and stretching for an estimated three thousand kilometres, these tunnels were incredibly difficult to film. They are for the most part stable, and will leave behind a record of our presence through anthroturbation (large-scale human tunnelling under the earth). Five mines operate in and around the city of Berezniki, collectively composing an underground web of an estimated ten thousand kilometres of tunnels. As a result, the town of Berezniki has experienced giant sinkholes that have swallowed roads and buildings and shut down the local railway station. Many residents have moved, despite the jobs available at the mines; there have even been calls to relocate the entire town.[1] The potash mined here is ultimately destined to fertilize large industrial farms, like those in the Imperial Valley in California (pages 88–91).

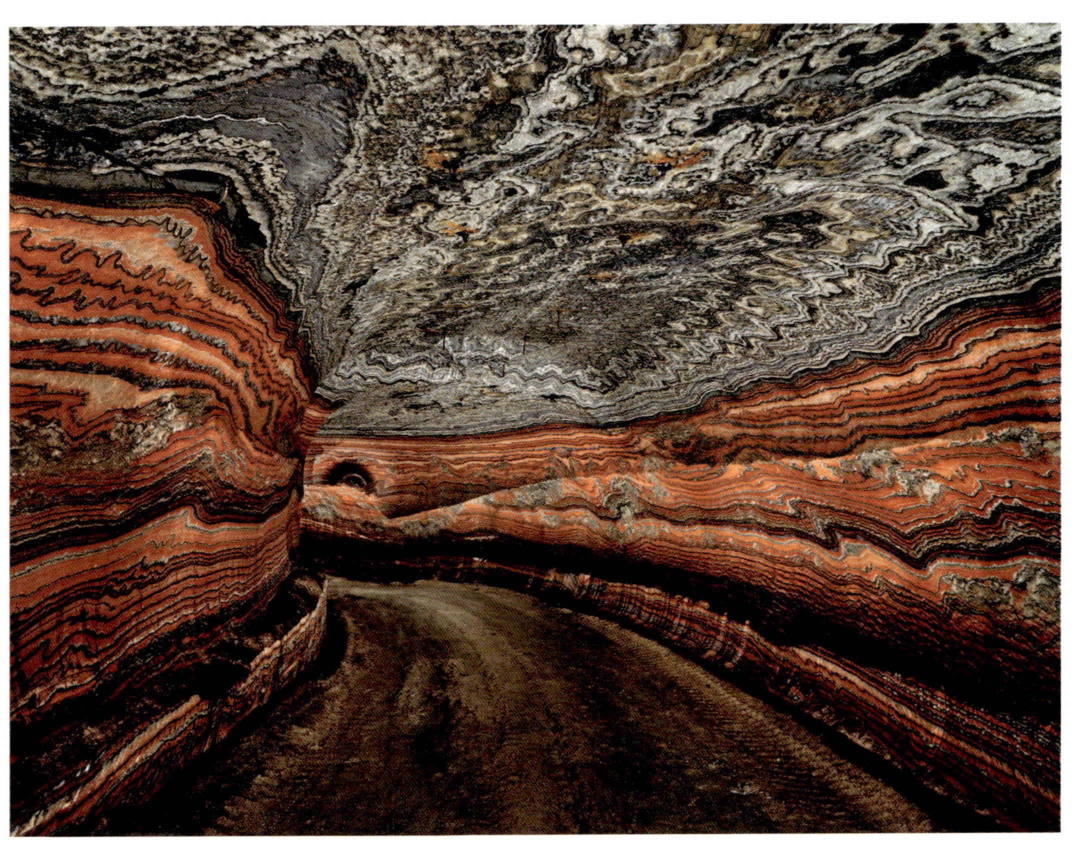

Edward Burtynsky, *Uralkali Potash Mine #2, Berezniki, Russia,* 2017

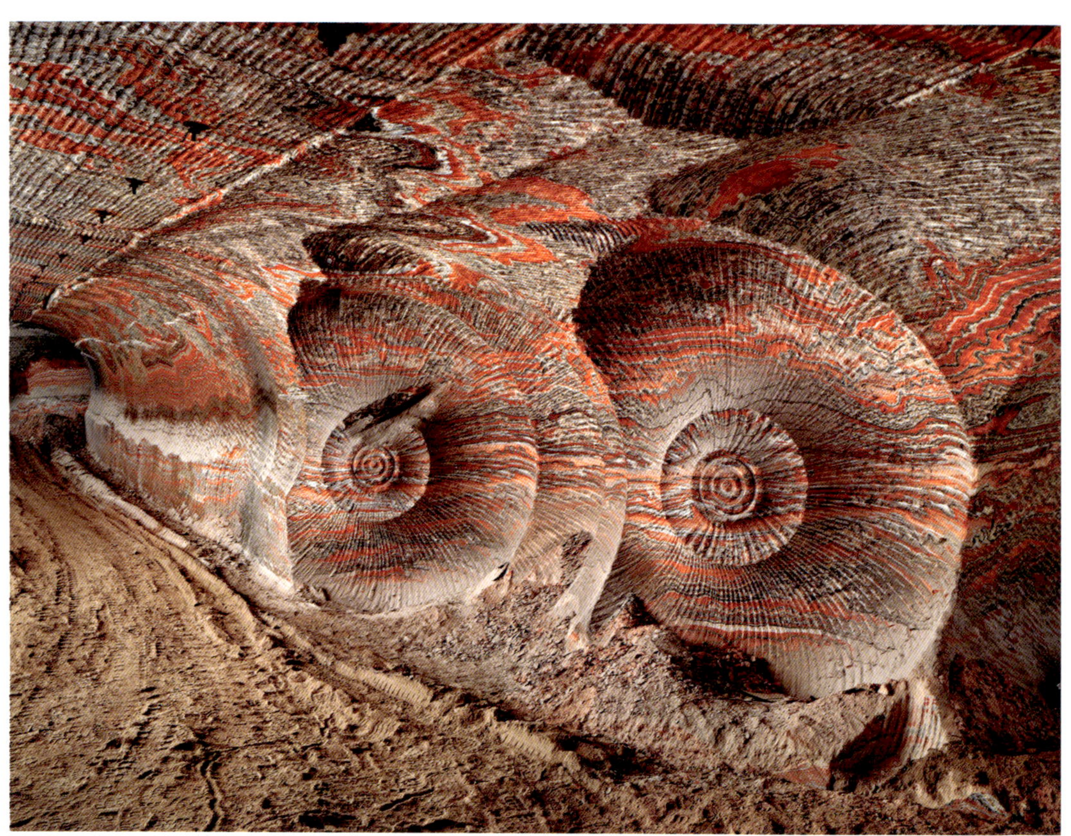

 Edward Burtynsky, *Uralkali Potash Mine #4, Berezniki, Russia*, 2017

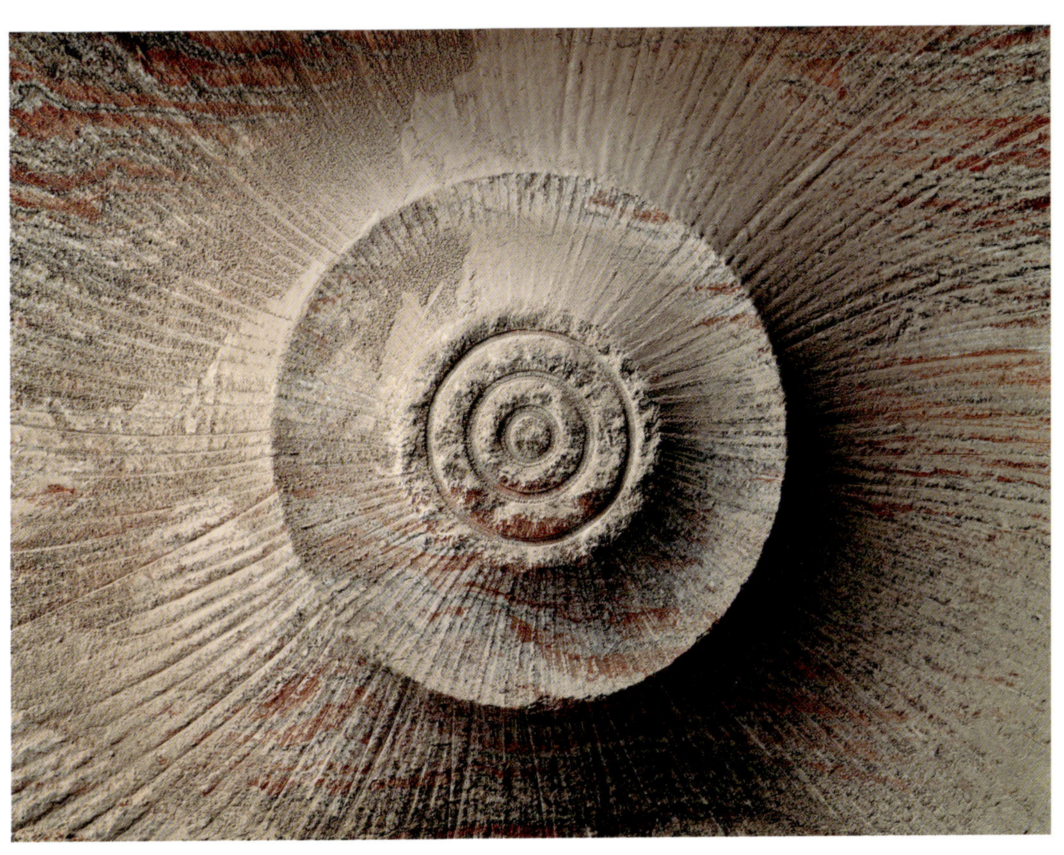

Edward Burtynsky, *Uralkali Potash Mine #6, Berezniki, Russia*, 2017

The Alps, formed over tens of millions of years by the collision of the African and Eurasian tectonic plates, are the highest and most extensive mountain range in Europe.[1] The mountains cover more than 200,000 square kilometres, creating a dense barrier between eight alpine countries, which, for many years, impeded long-distance travel in the region.[2] On June 1, 2016, the passage between northern and southern Europe became significantly swifter with the opening of the Gotthard Base Tunnel. Some fifty-seven kilometres in length, it is now the world's longest railway tunnel, and services full passenger and freight needs. In order to connect Switzerland's German- and Italian-speaking regions, the tunnellers had to bore through a wide range of rock strata, ranging from solid granite to sedimentary rock. Approximately 28.3 million tonnes of material were excavated during construction.[3] With the rock overhead measuring up to 2,300 metres in depth, the Gotthard Base Tunnel is also the deepest railway tunnel in the world.[4] Gotthard is an example of anthroturbation or human tunnelling, a significant category of Anthropocene Working Group research. For this installation, we strapped our RED Epic camera to the front of the train and filmed in 5K resolution for the entire journey through the tunnel, which took approximately twenty minutes.

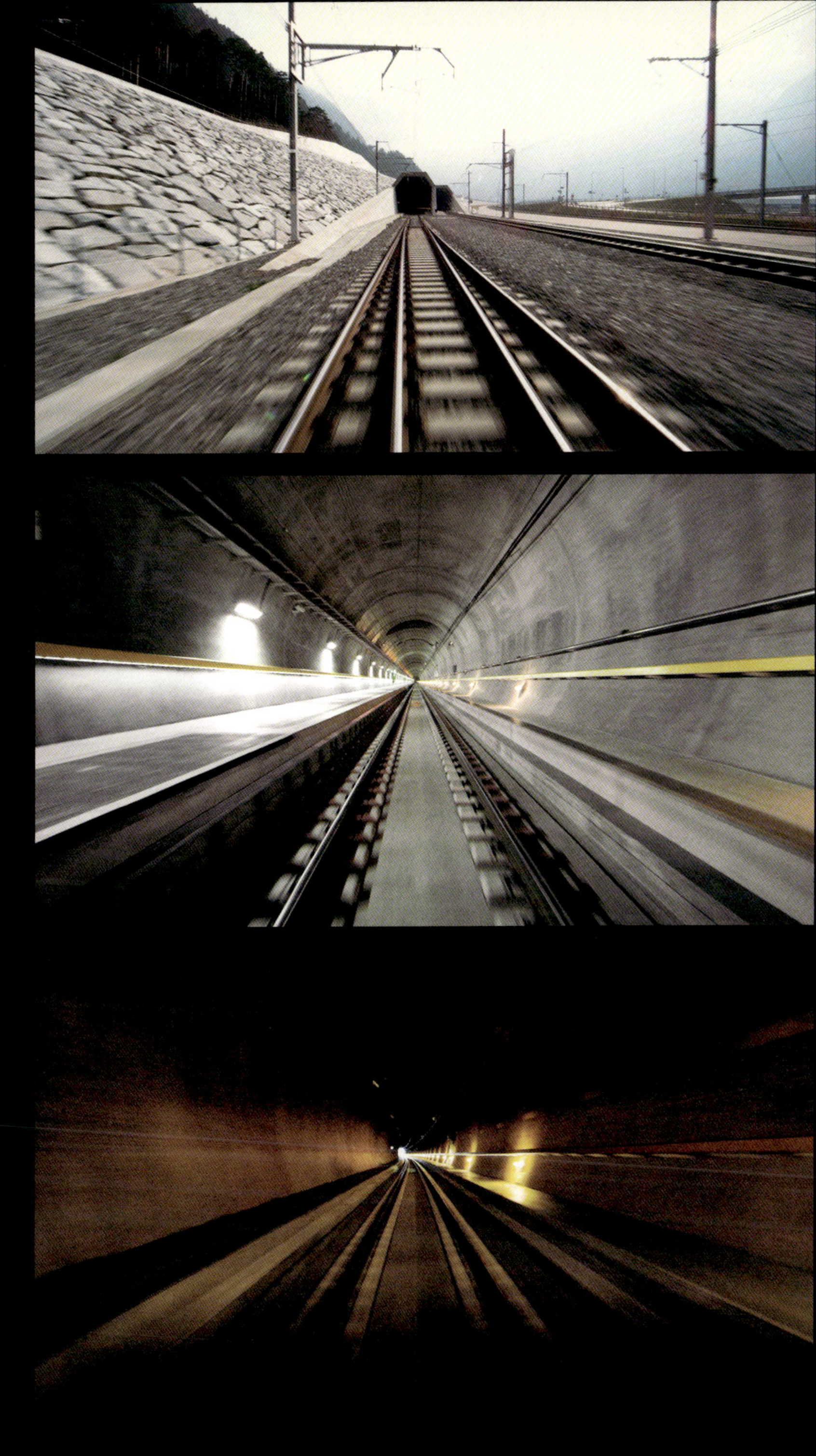

Jennifer Baichwal and Nicholas de Pencier, *Gotthard Base Tunnel, Gotthard, Switzerland* (film stills), 2018

Marble production is one of the most important sectors in Italy, its yield accounting for 18 percent of the world's output.[1] The marble quarries in Carrara have been mined since the time of Ancient Rome. This stone was famously used by Michelangelo (his *David* statue is made from a single block from Carrara), who would stay for months at a time to supervise its removal. Until the sixteenth century, this work was carried out by slaves who used metal chisels and wooden wedges, expanded by water and placed in the natural cracks of the stone, to separate the marble blocks from the mountain.[2] Extraction was sped up exponentially in the eighteenth century with the arrival of explosives that, although efficient, left behind huge piles of waste material locally referred to as *ravaneti*.[3] By this time, industrial processing in the region began to increase, thus creating important factories established for the cutting and polishing of slabs.[4] Beginning in the 1960s, the introduction of trucks and excavators extended transport capacity, while helical wire, used to precisely cut the marble from the quarry in place of wasteful explosives, reduced incision times.[5] Asked how much marble was left in the mountain, a quarry owner said, "In my sixty-three years of working here, it's as if I only plucked a hair from a pig." These mountains have been worked continuously for more than three millennia, and to this day still appear near limitless in their capacity to give up marble. The negative architecture formed on the land by quarries operating at this scale leaves its lasting trace upon our planet and is visible from space.

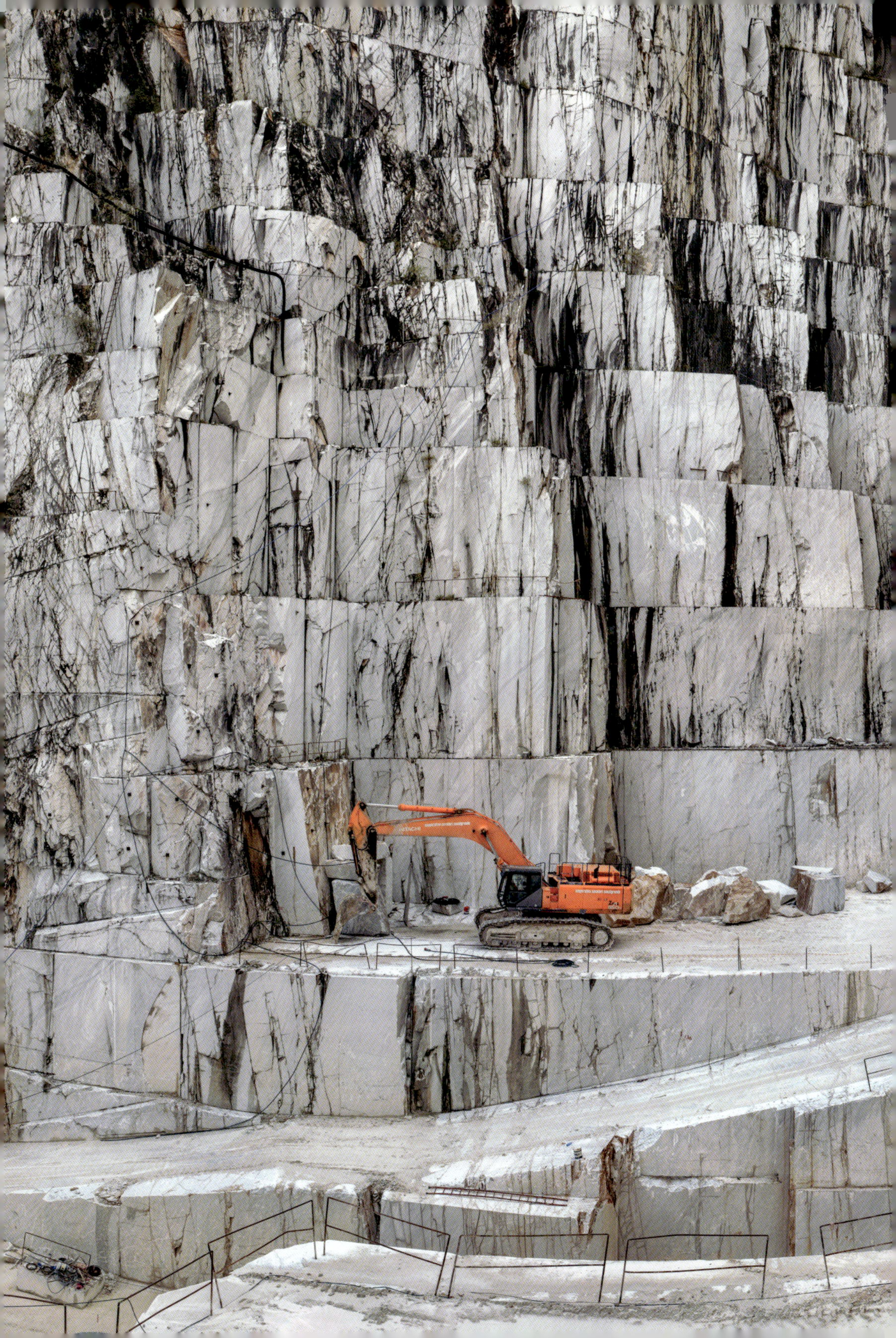

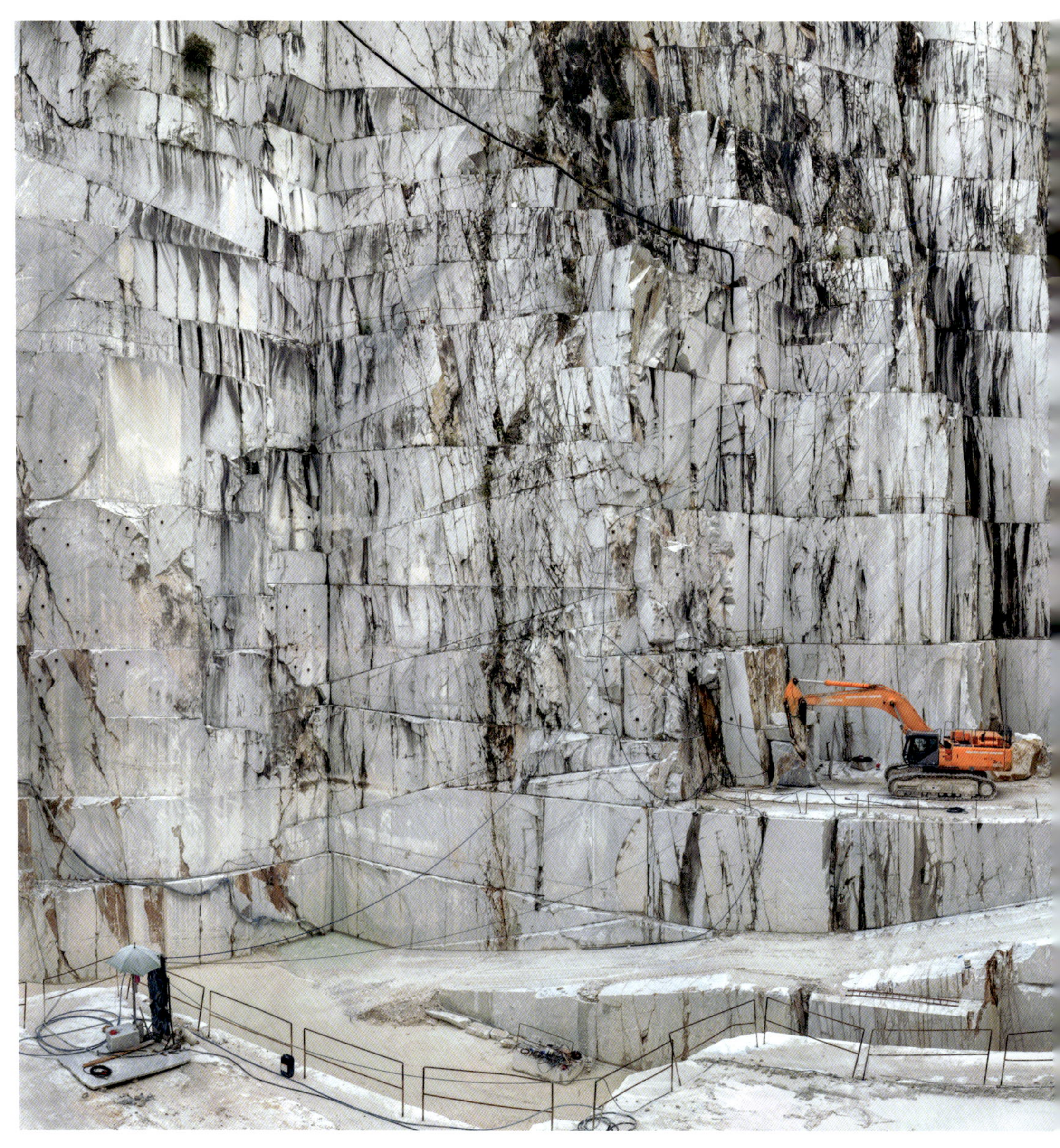

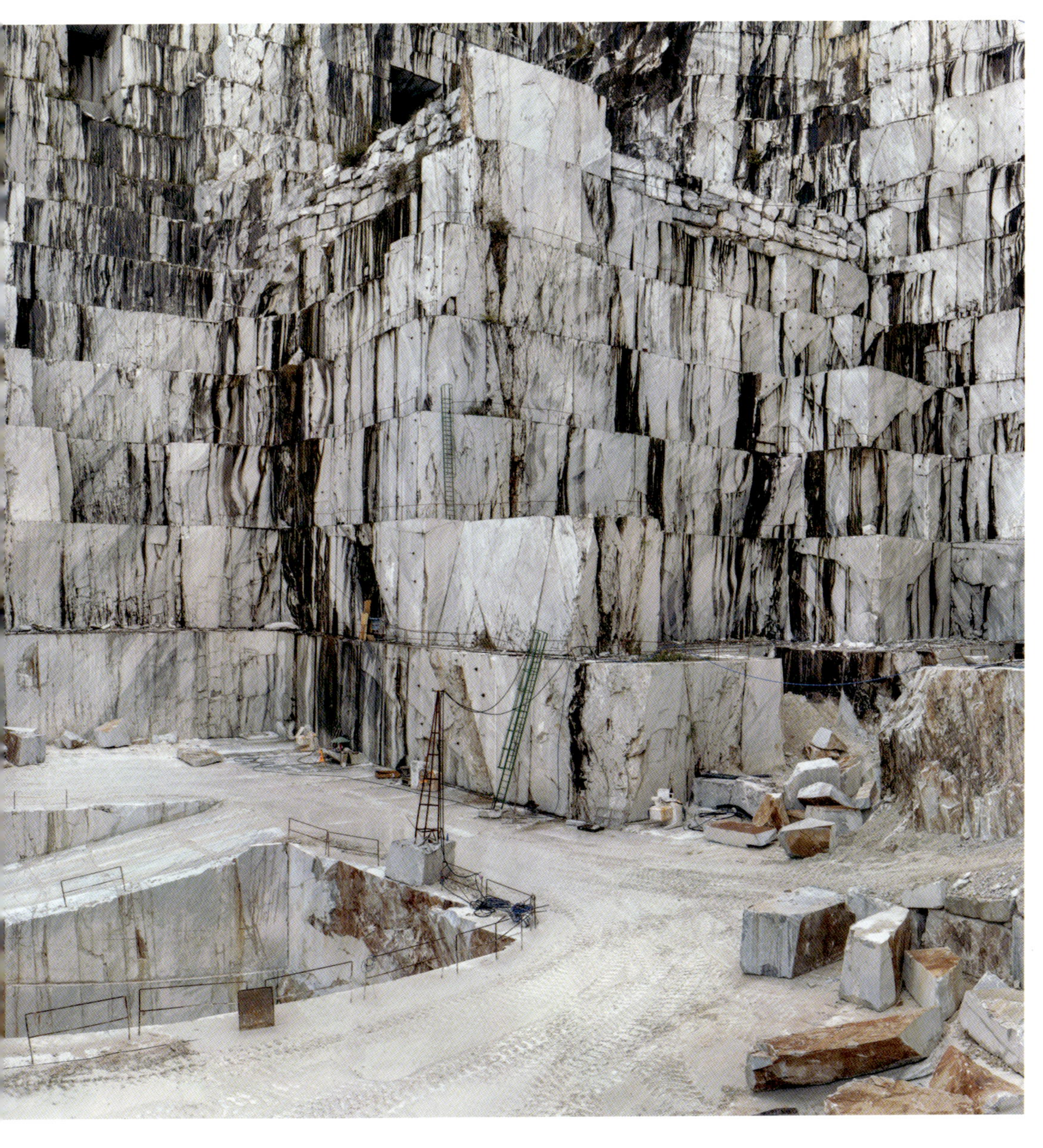

Edward Burtynsky, *Carrara Marble Quarries, Cava di Canalgrande #2, Carrara, Italy,* 2016

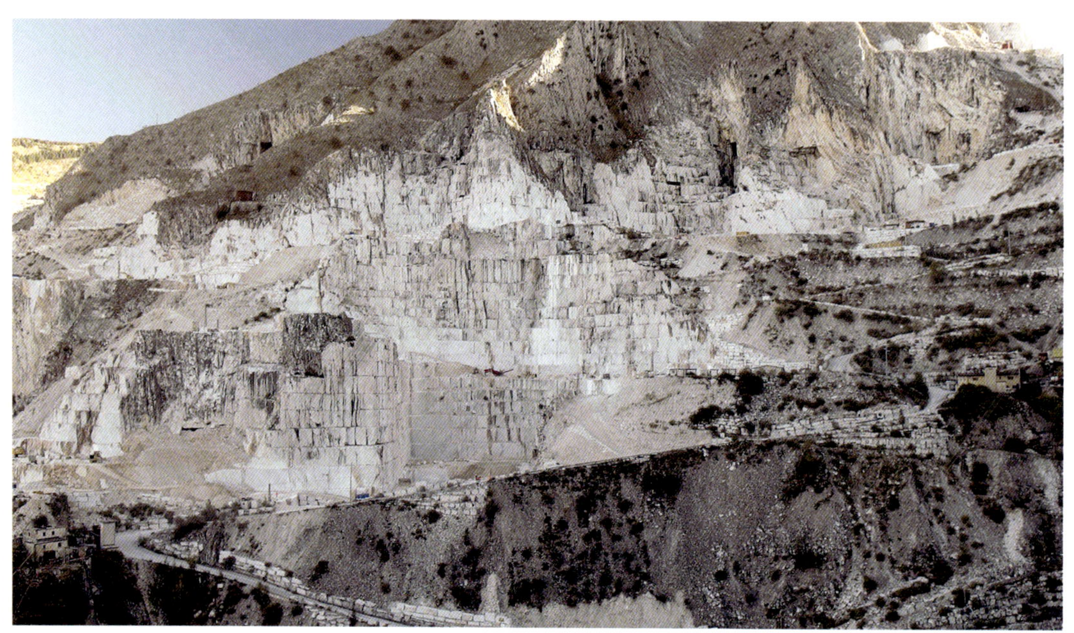

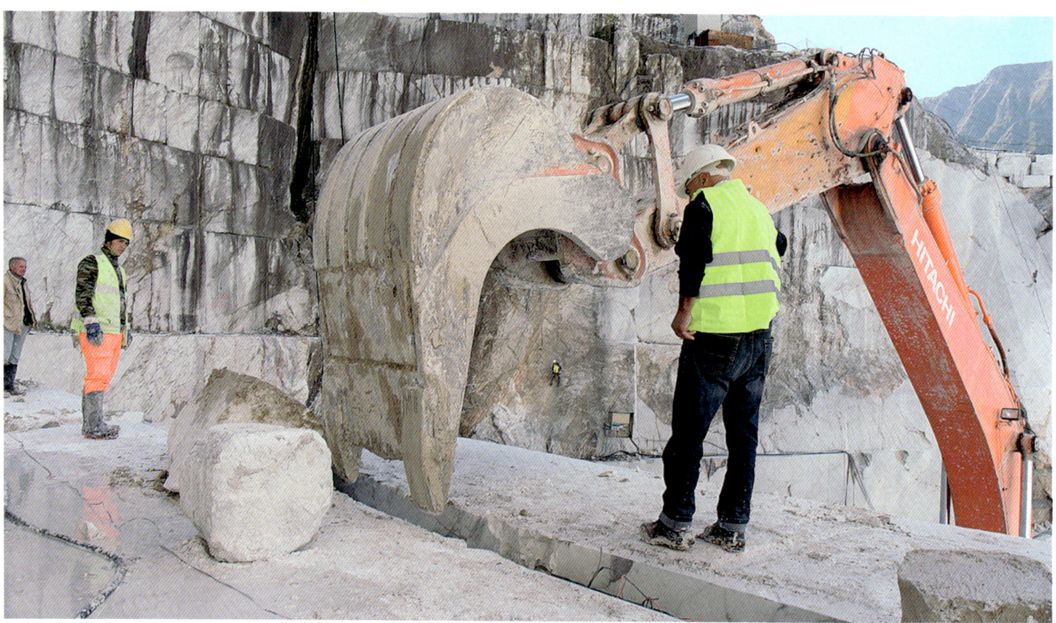

Jennifer Baichwal and Nicholas de Pencier
Falling Slabs, Carrara, Italy, 2018
Cava di Canalgrande, Carrara, Italy, 2018

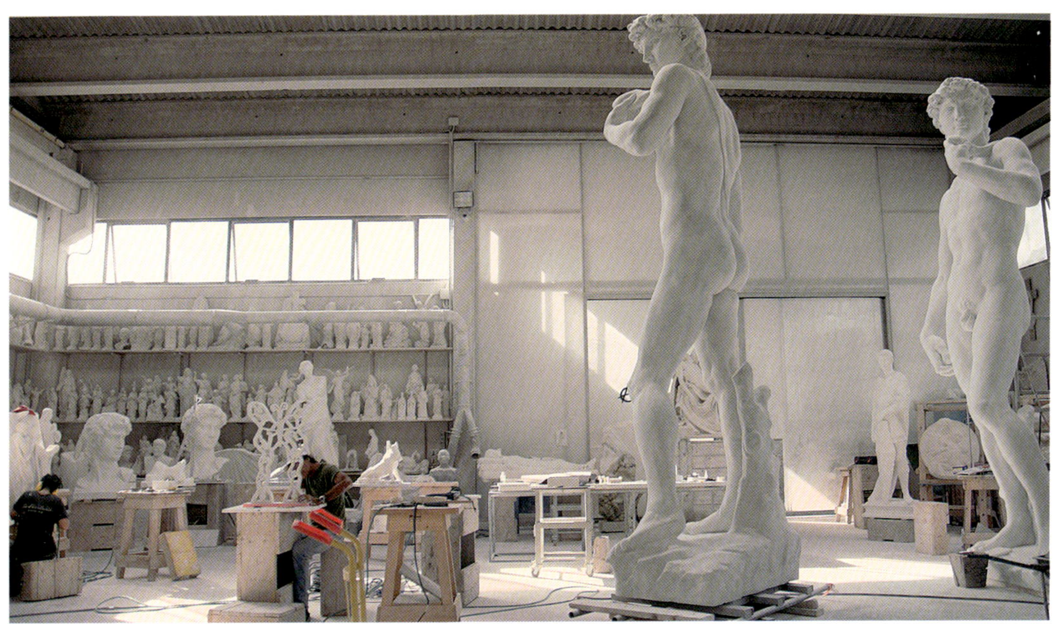

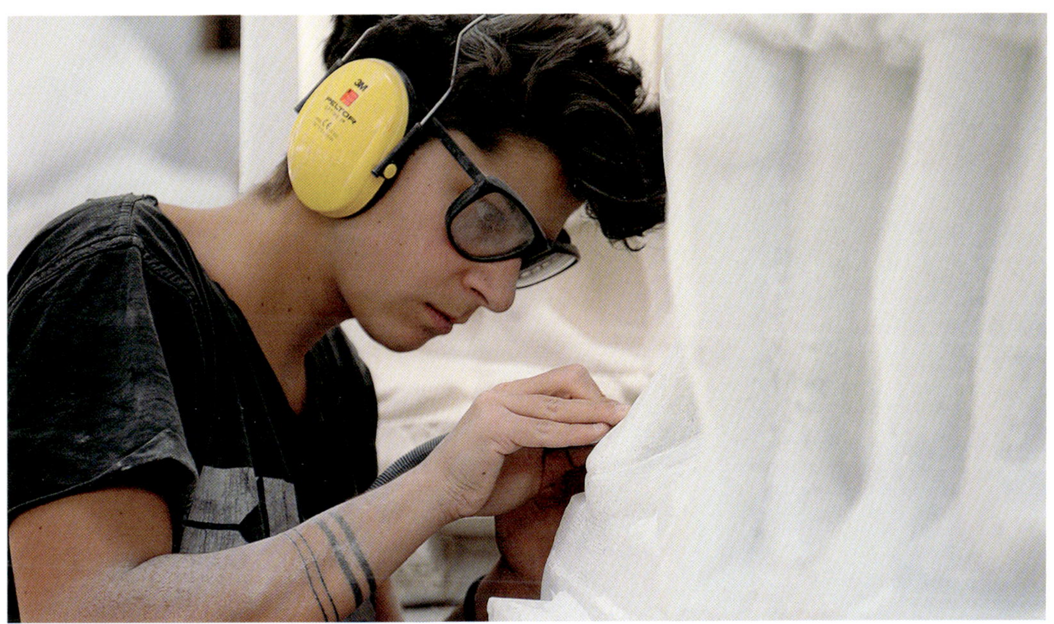

Jennifer Baichwal and Nicholas de Pencier
Carving Studio, Carrara, Italy, 2018

In 2015, global power consumption totalled 575 quadrillion British thermal units (BTUs), or over 168,515 terawatts.[1] In 2013, just over 30 percent of all global greenhouse gas emissions came from electricity and heat production, excluding industry.[2] By the end of 2017, the global wind power capacity was estimated at 539,291 megawatts (or 539 gigawatts), representing 5 percent of global electricity demand.[3] The cost of wind power is falling rapidly, similar to what is happening with solar power. These wind farms represent an early advance of an industry that, in combination with solar, will see three-quarters of all new power energy investment by the mid-twenty-first century.[4]

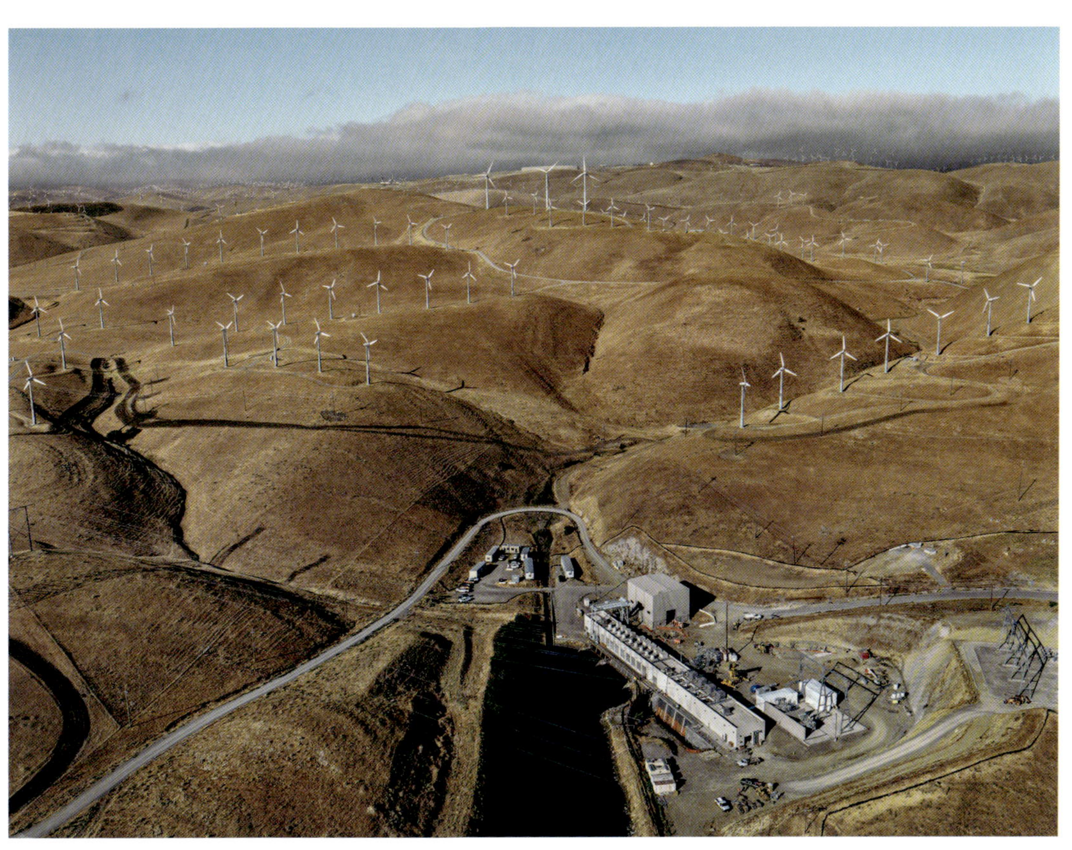

Edward Burtynsky, *South Bay Pumping Plant #1, Near Livermore, California, USA*, 2009

The Atacama Desert is one of the world's most optimal places for solar energy. Perched at an altitude of 1,500 metres above sea level, the desert here is subject to a sun that burns bright and hot, hampered by only two or three cloudy days per year. When completed, the solar farm seen here, Atacama-1, will use a 2,000-tonne solar receiver to collect the heat from 10,600 heliostatic mirrors that follow the sun like flowers.[1] Heat is retained in a 50,000-tonne pool of molten salt that drives turbines through the night, making this system viable for twenty-four-hour power generation.[2] The salt is locally mined, as it is in the nearby lithium flats (pages 106–107). Most of the energy that is generated in facilities like this is sold to industry, and a significant portion powers the Chuquicamata copper mine (pages 134–137).[3] The very first solar-thermal towers ever built were PS10 and PS20 in Spain, which opened in 2007 and 2009 respectively. Functionally, they are very similar to the Atacama-1 tower, though they lack the liquid salt to store heat.[4] Capital costs for all types of solar installations fell a dramatic 73 percent between 2010 and 2017.[5] In 2016, solar represented about 47 percent of newly installed renewable power capacity, with wind and hydropower accounting for 34 percent and 15.5 percent respectively.[6] Shifting priorities in energy investment, as well as international political efforts to control climate change, suggest that the geopolitical incentives to move away from fossil fuels may become even stronger.

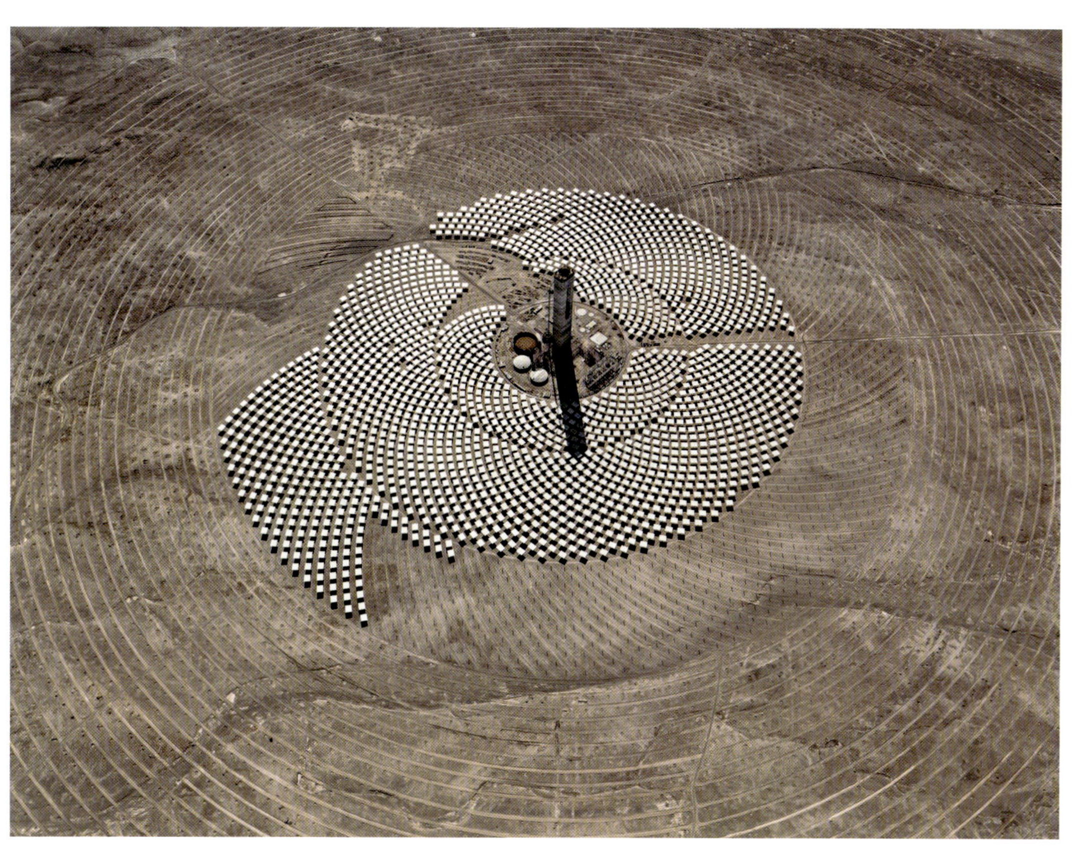

Edward Burtynsky, *Cerro Dominador Solar Project #1, Atacama Desert, Chile,* 2017

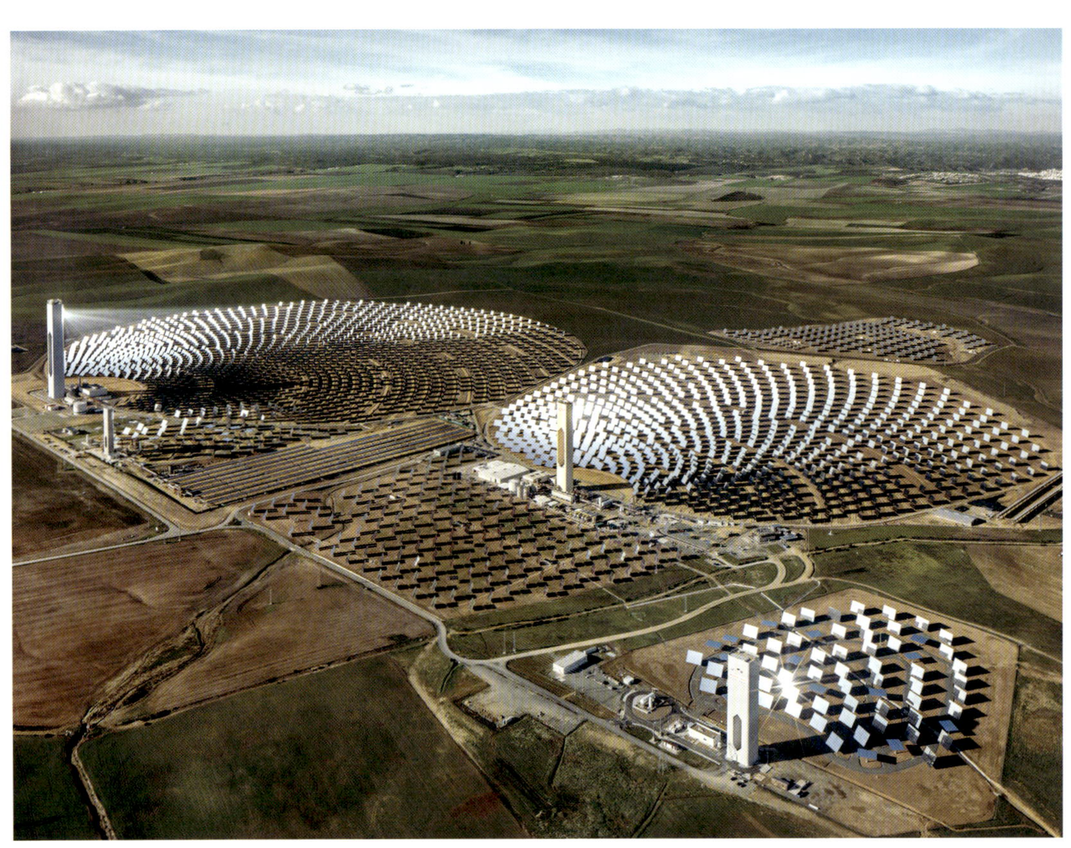

 Edward Burtynsky, *PS10 Solar Power Plant, Seville, Spain,* 2013

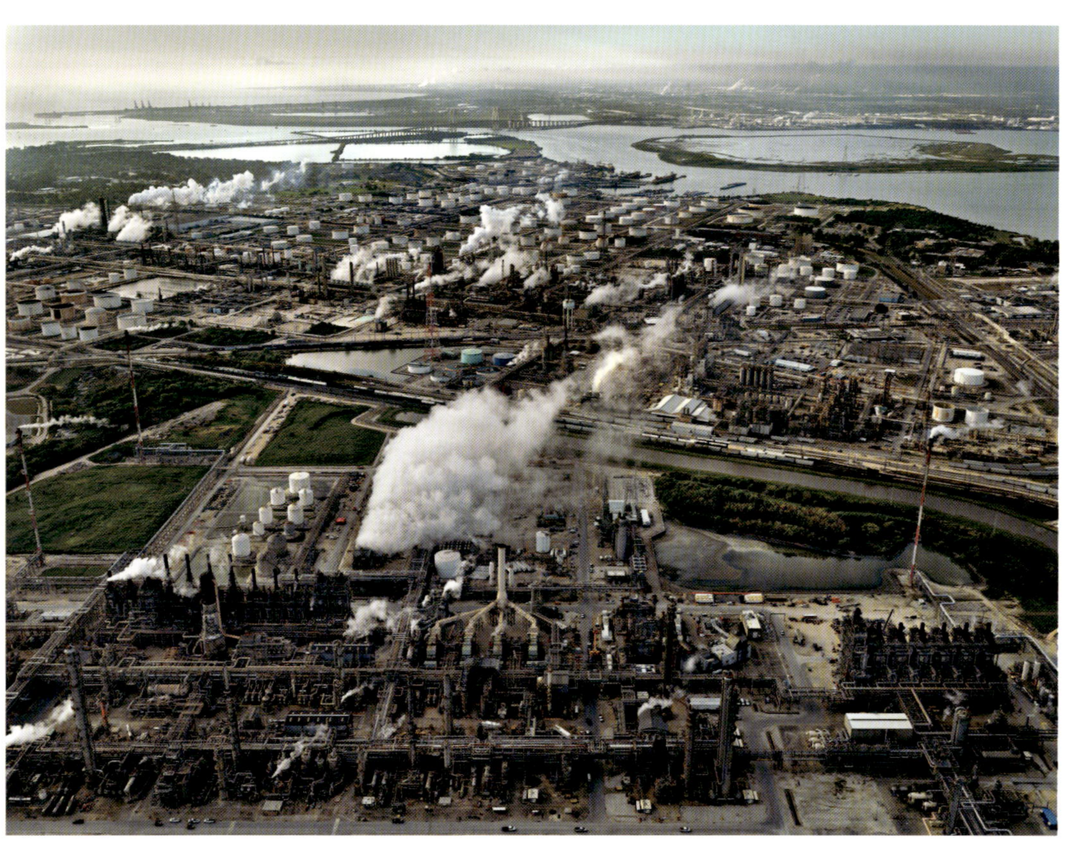

Edward Burtynsky, *Petrochemical Plants, Baytown, Texas, USA*, 2017

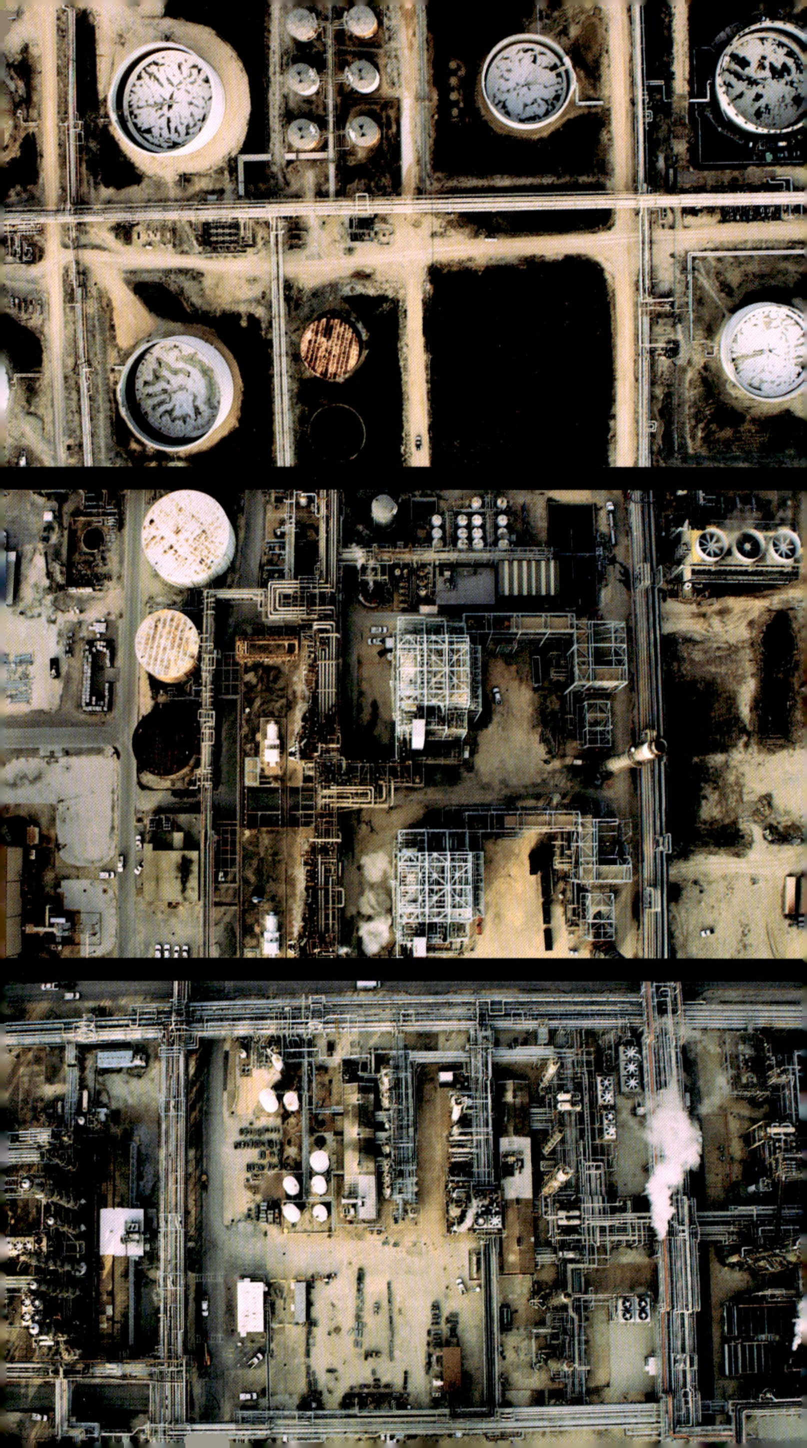

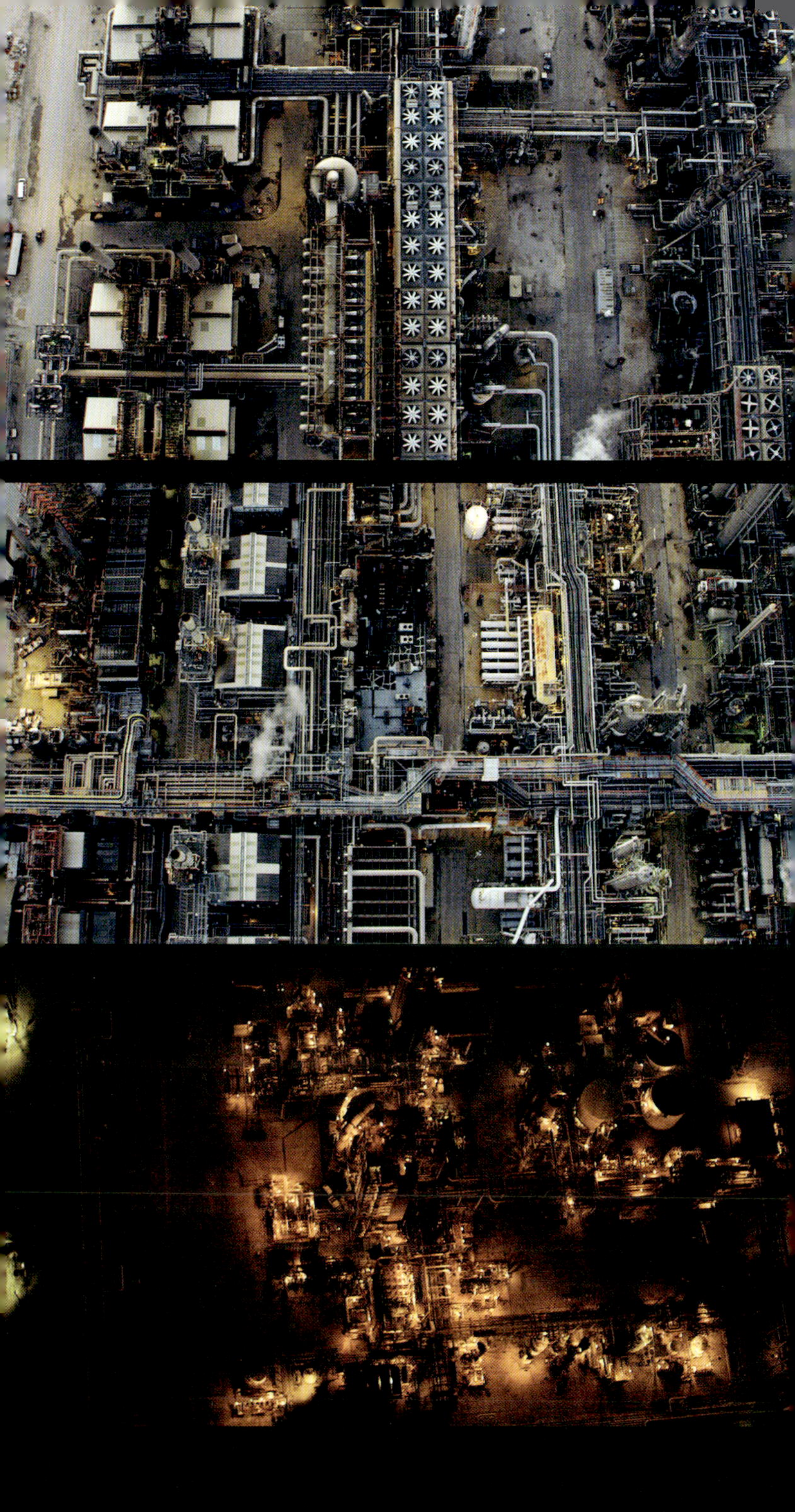

Jennifer Baichwal and Nicholas de Pencier, *Oil Refineries, Texas, USA* (film stills), 2018

Many refineries and pumping stations, like the one pictured here in Long Beach, California, are in the centre of major metropolitan areas; petrochemical products are integrated into every aspect of our lives and are therefore hard to replace. With no ready alternative natural resource to replace petrochemicals, finding a sustainable substitute for crude oil and all its complex carbon chains continues to prove a difficult task. Powerful corporate interests that exercise regulatory capture make shifting to alternative energies far more difficult for democratic societies to achieve.

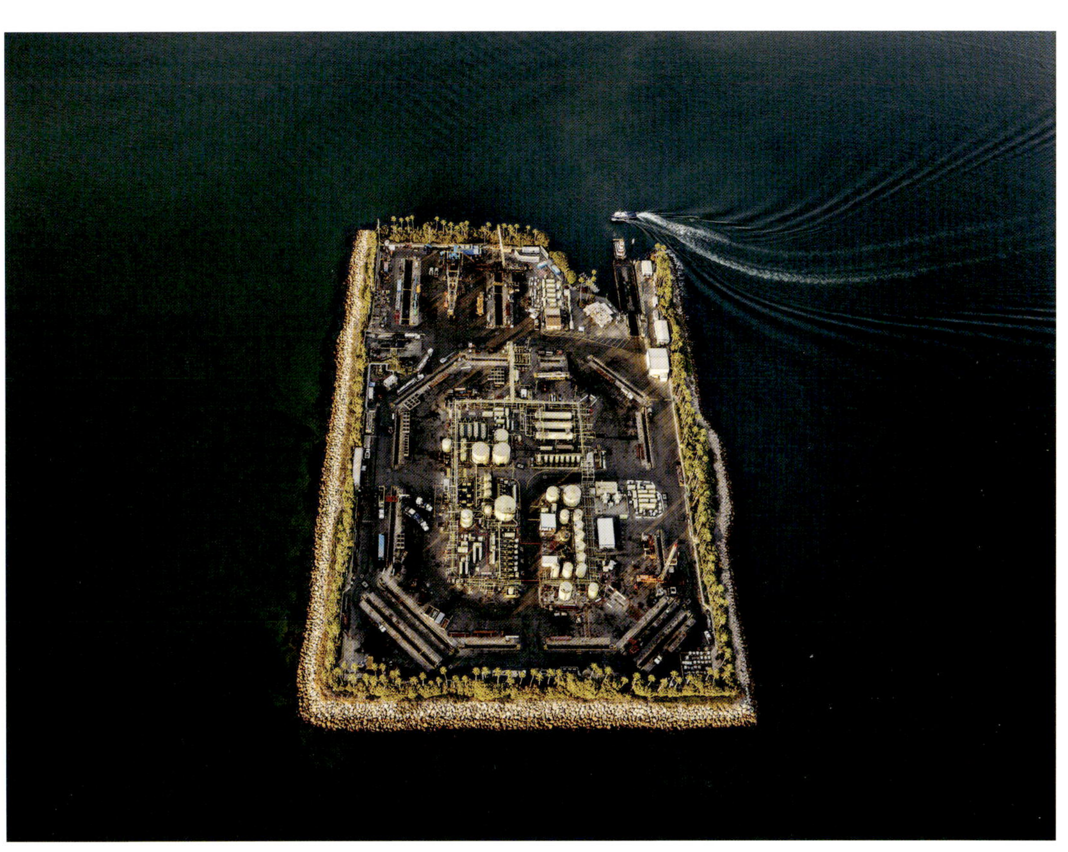

Edward Burtynsky, *Freeman Island, Long Beach, California, USA*, 2017

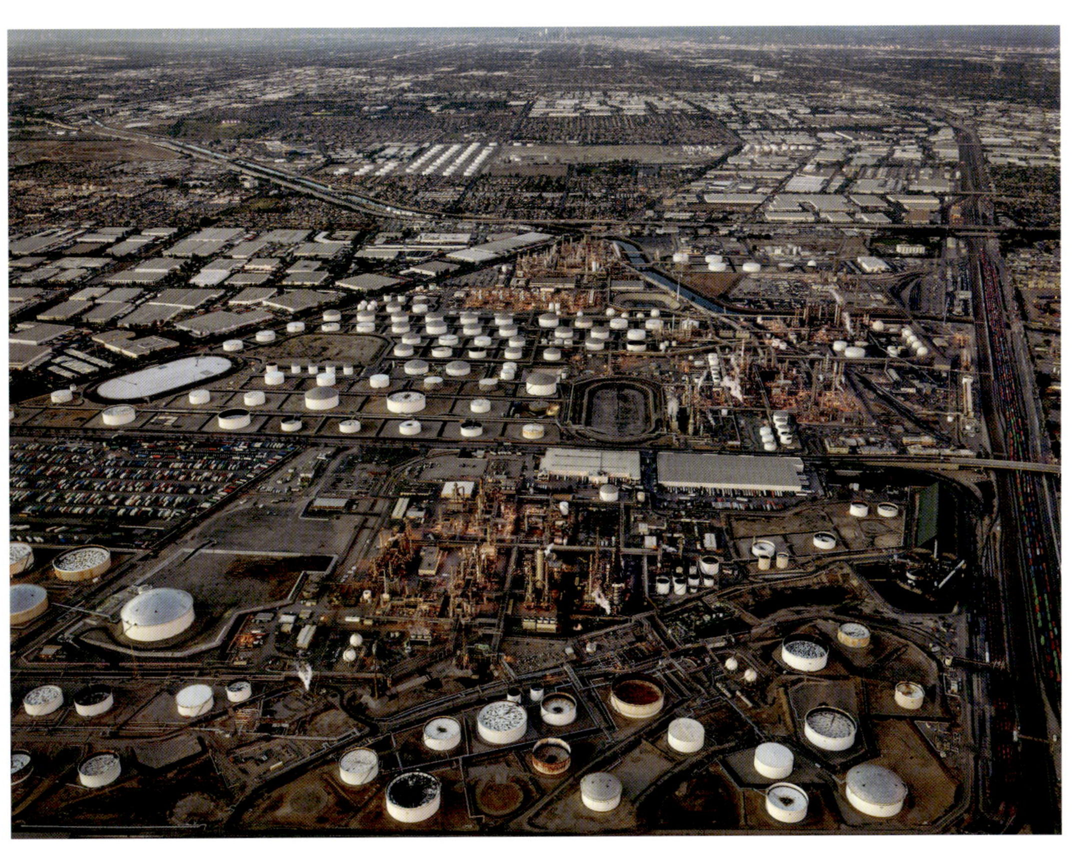

Edward Burtynsky, *Oil Refineries and Storage, Carson, California, USA*, 2017

These images depict great rainforests under stress. The temperate rainforests of the Pacific Northwest coast are an incredible site of biodiversity. British Columbia, whose rainforests are pictured here, occupies only 10 percent of Canada's geographical area, yet the province contains more than half of Canada's vertebrates and vascular plants, as well as three-quarters of its bird and mammal species.[1] Some of the tallest trees in the world can be found in the old-growth forests of this region. Commercial logging on Vancouver Island dates back to the 1820s, with the first sawmills being established in the 1860s.[2] The introduction of the combustion engine in the 1940s saw previously unreachable forested areas turned into logging sites, their timber then easily transported by road.[3] Today, 90 percent of B.C.'s logging occurs on publicly owned Crown lands.[4] While less than 1 percent of the province's forests are harvested annually,[5] Vancouver Island's primary rainforests are logged at three times the rate of tropical regions.[6] As of the early twenty-first century, only 10 percent of Vancouver Island's old-growth forests remained, and logging of these spaces continues.[7] By 2017, merchantable pine from the B.C. Interior was reduced by more than half due to mountain pine beetle infestations.[8] As climate change encourages further infestations, among other impacts, logging companies will increasingly focus on exports from the coasts.[9] Like those at the saw mills beside Makoko in Nigeria, log booms are used to transport raw timber from the area where it was harvested to central log yards (pages 97, 104–105). In both cases, increasingly globalized economies have led to mass export of local resources. In the case of British Columbia, from 1990 to 2014 more than half of the province's sawmills closed due to offshore outsourcing of processing.[10] Between 2013 and 2016 nearly 26 million cubic metres of raw logs were shipped from B.C. without value-added local processing. Research by the Ancient Forest Alliance shows that exports have also made it affordable to harvest in more remote areas of forest.[11] Like the vanishing primary forests of Borneo, the old-growth covered areas of British Columbia provide important carbon sequestration, at rates significantly higher than secondary or degraded forests. And while a visitor to Cathedral Grove on Vancouver Island may be struck by the feeling that they are in an eternally regenerating wood, they would be wise to note that such forests are as vulnerable as they are majestic.

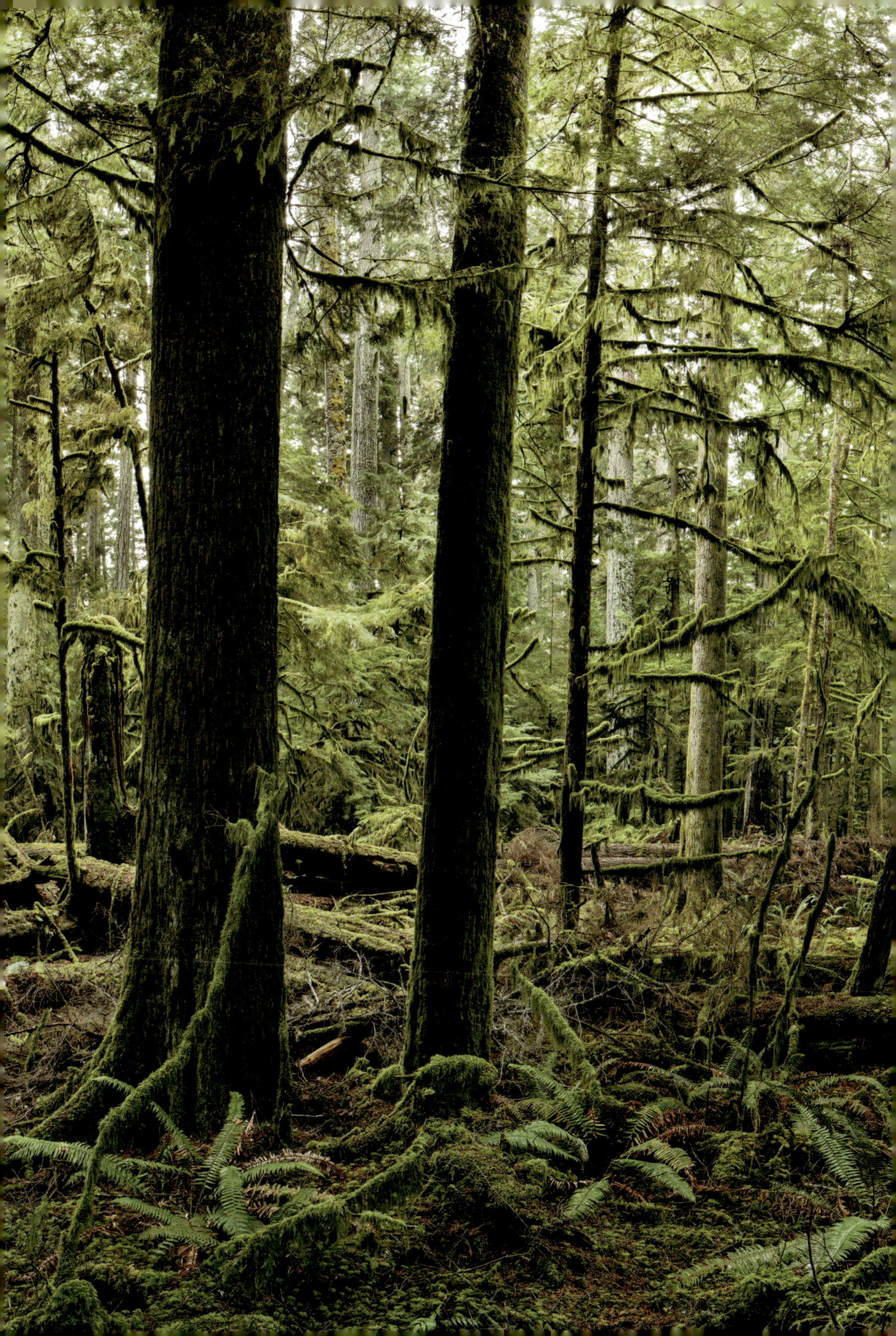

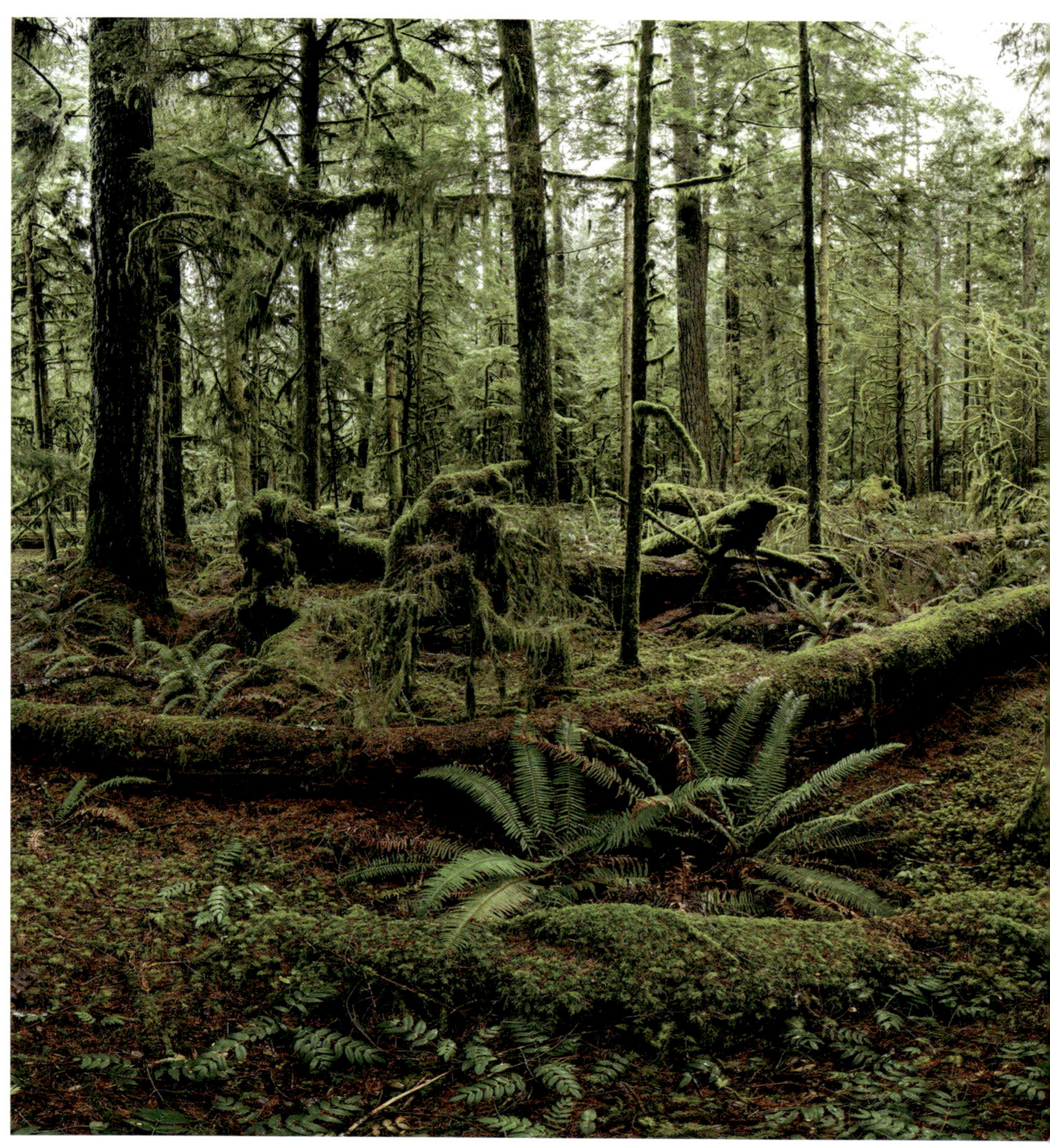

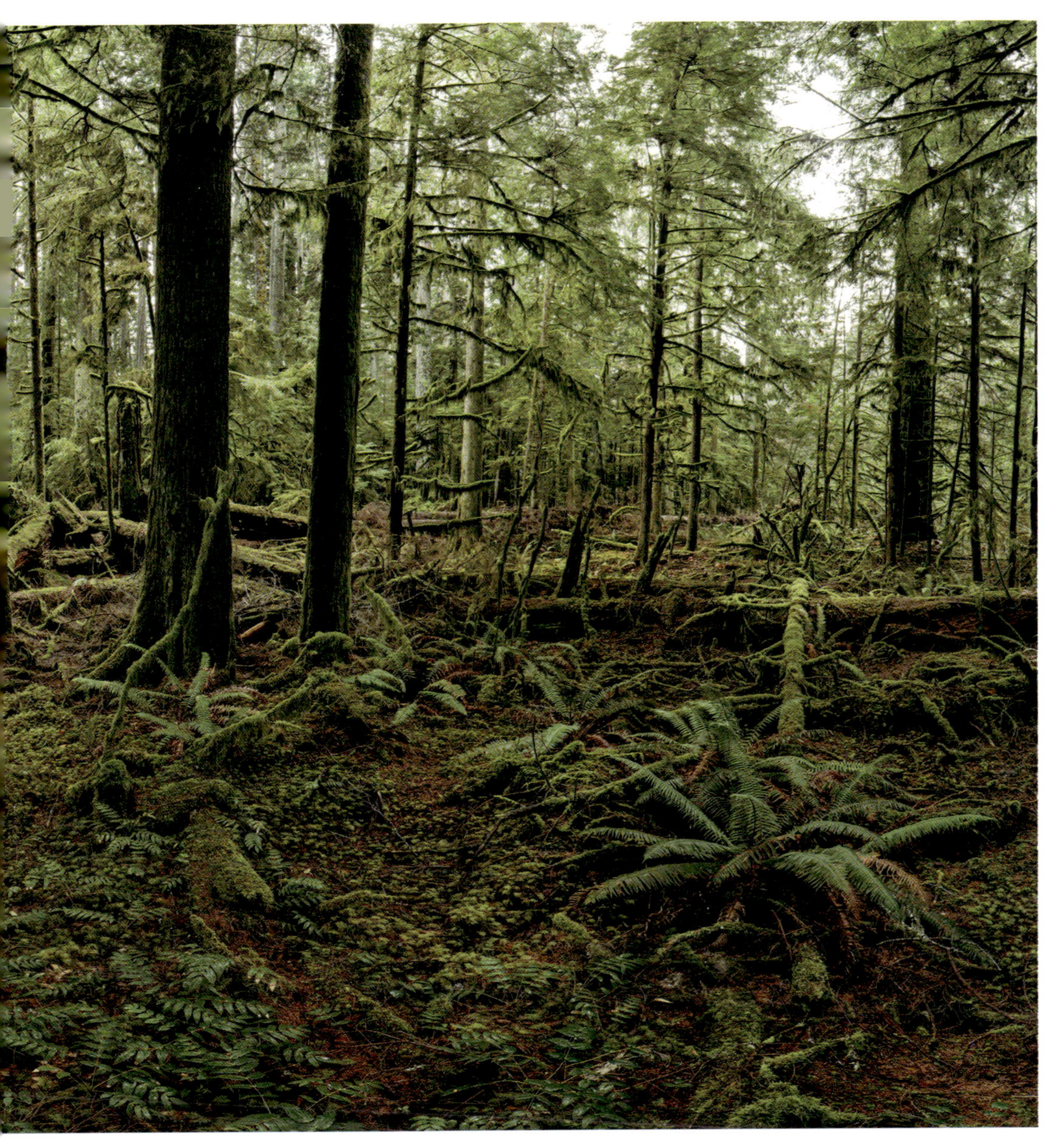

Edward Burtynsky, *Cathedral Grove #1, Vancouver Island, British Columbia, Canada,* 2017–2018

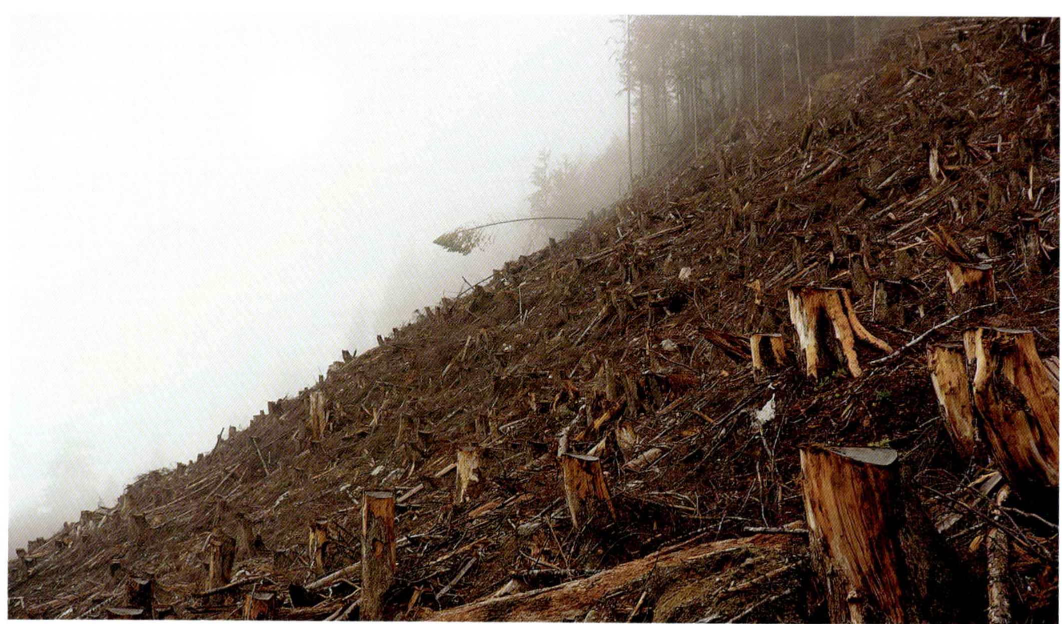

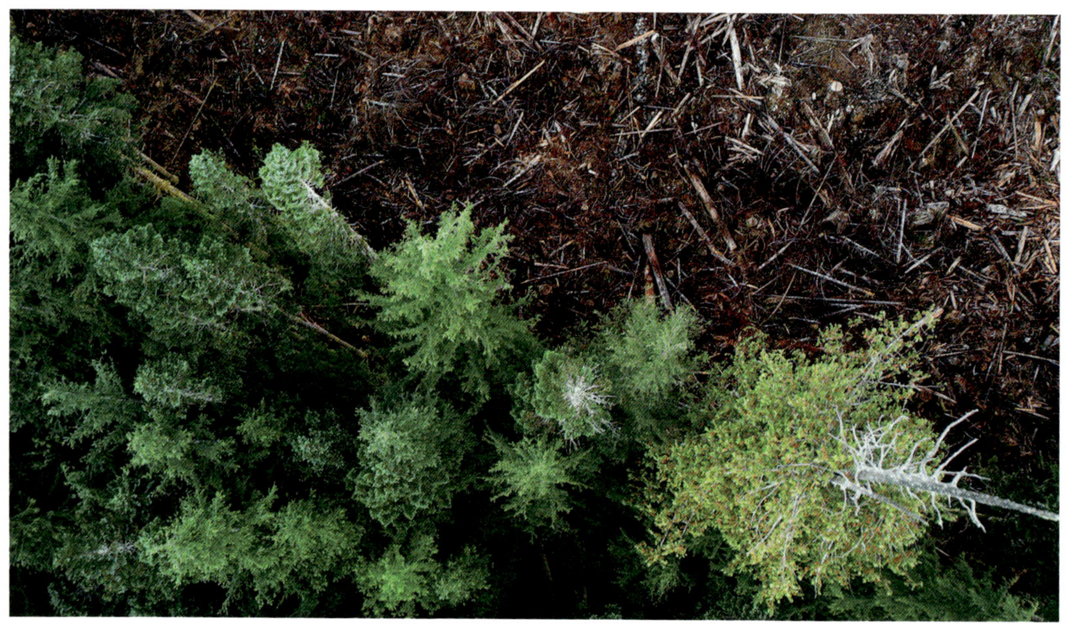

Jennifer Baichwal and Nicholas de Pencier
Clearcuts, Vancouver Island, Canada, 2018

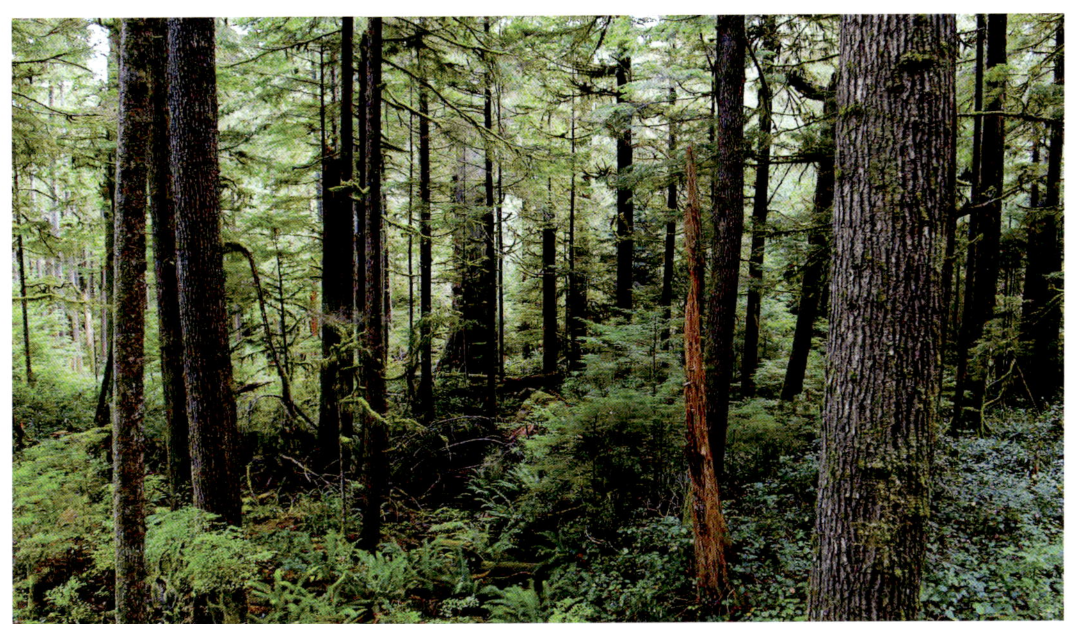

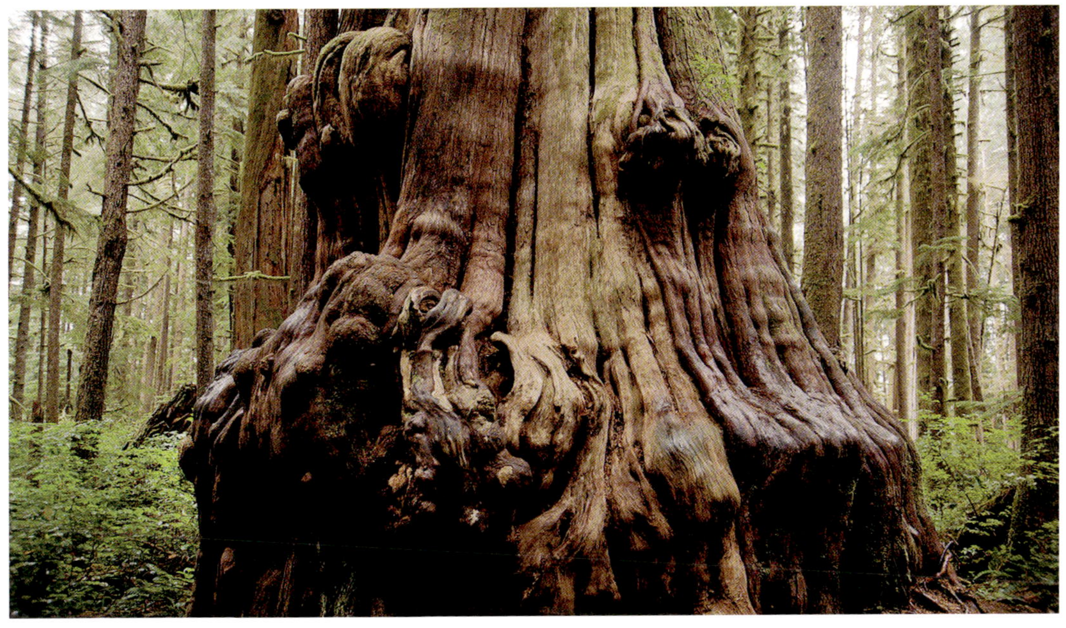

Jennifer Baichwal and Nicholas de Pencier
Tree Felling, Vancouver Island, Canada, 2018
Old-Growth Tree Portraits, Vancouver Island, Canada, 2018

This coral wall in Indonesia is a rare remnant of our globally diminishing coral reefs. In Komodo, unique local conditions allow reefs to flourish despite the corrosive forces of climate change. This ideal habitat is created through the convergence of strong daily tidal flows and nutrient-rich waters from the Indian Ocean.[1] This pristine reef supports abundant life—fish, corals, crustaceans, cartilaginous aquatic creatures (mantas), marine reptiles (sea turtles), and mammals (dolphins). These dynamic forces made this environment a difficult one to photograph. Shot in multiple frames and stitched together later, the final result is a vibrant document of an increasingly rare ecosystem. Indeed, coral bleaching may be more likely to occur here (as elsewhere) should sea water temperatures begin to rise.[2] In 2016, the Great Barrier Reef in northeastern Australia, the world's largest reef system, suffered a devastating mass bleaching event (page 176). In all, about 22 percent of the entire reef's corals were lost.[3] Further diminishing hope for recovery, a second wave of mass bleaching was recorded in 2017.[4] Bleaching is not limited to Australia, and is seen on other coasts with increasing frequency as ocean temperatures rise and acidity increases.[5] Australia is the world's largest coal exporter, and many of the mines and ports are centred near Queensland, where reefs experienced the greatest damage.[6] The tension here between conservation and industry has been lengthy and ongoing. While there is hope that the coral reefs will recover, the increased frequency of bleaching events will only put more stress on those that survive.

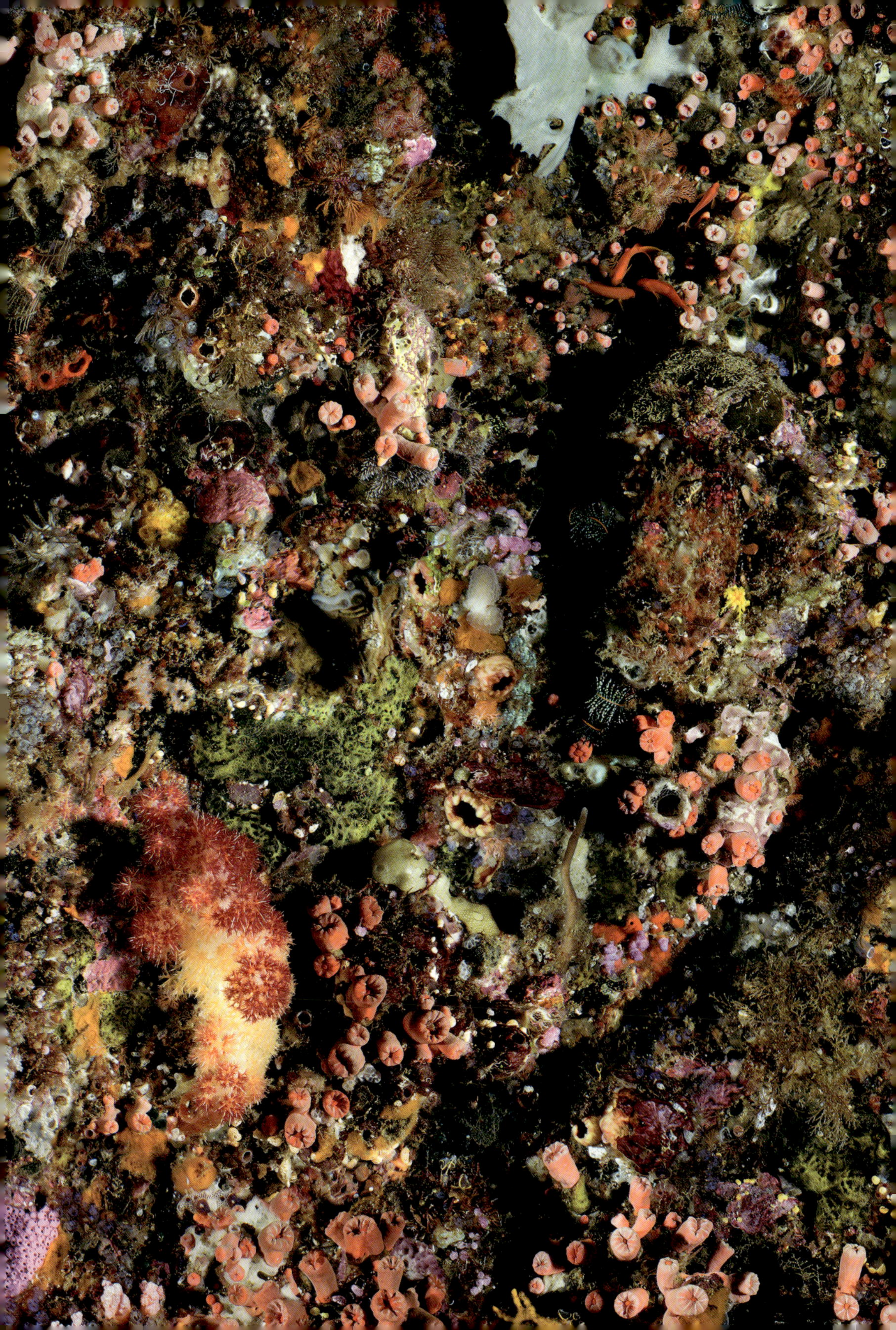

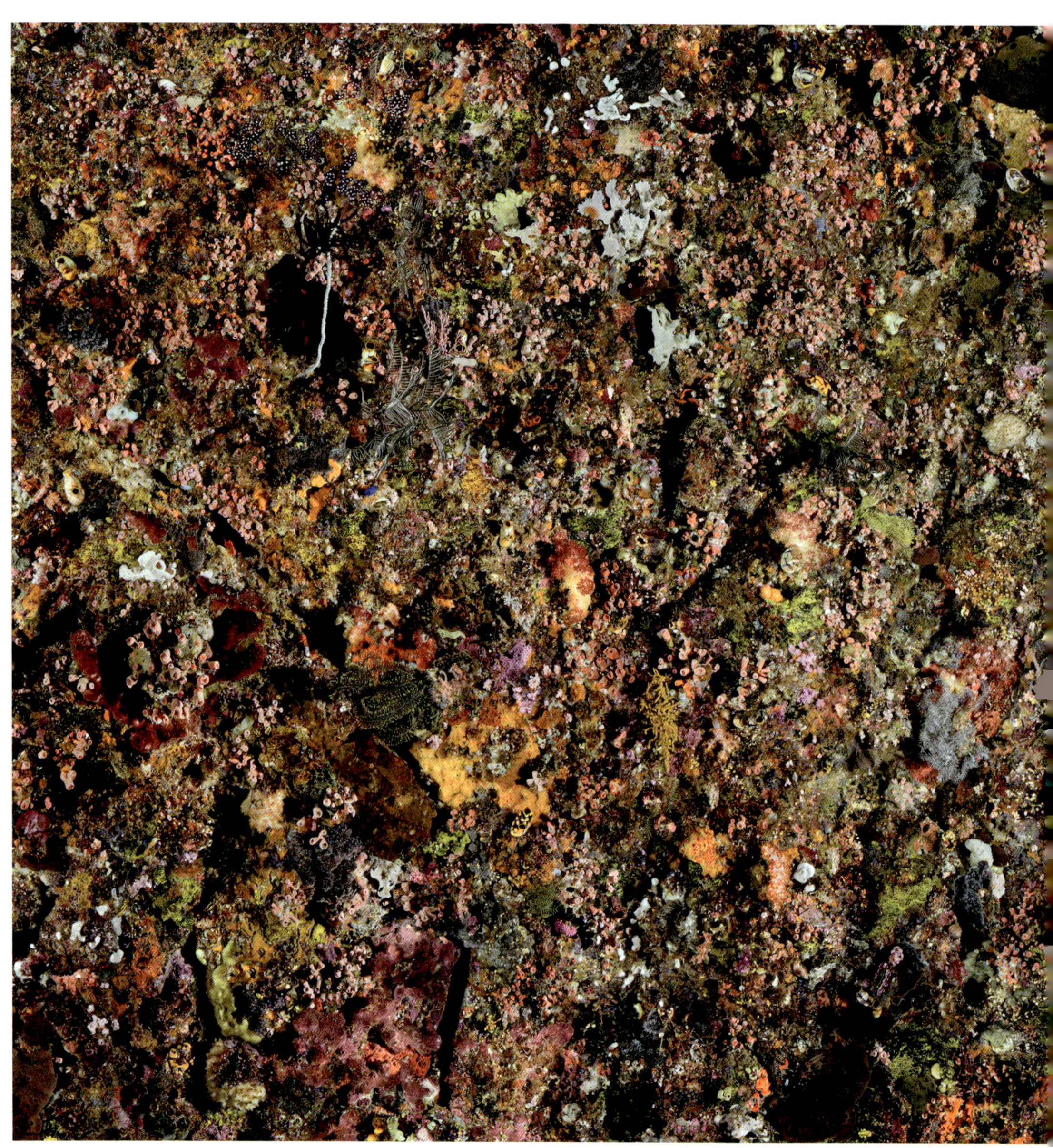

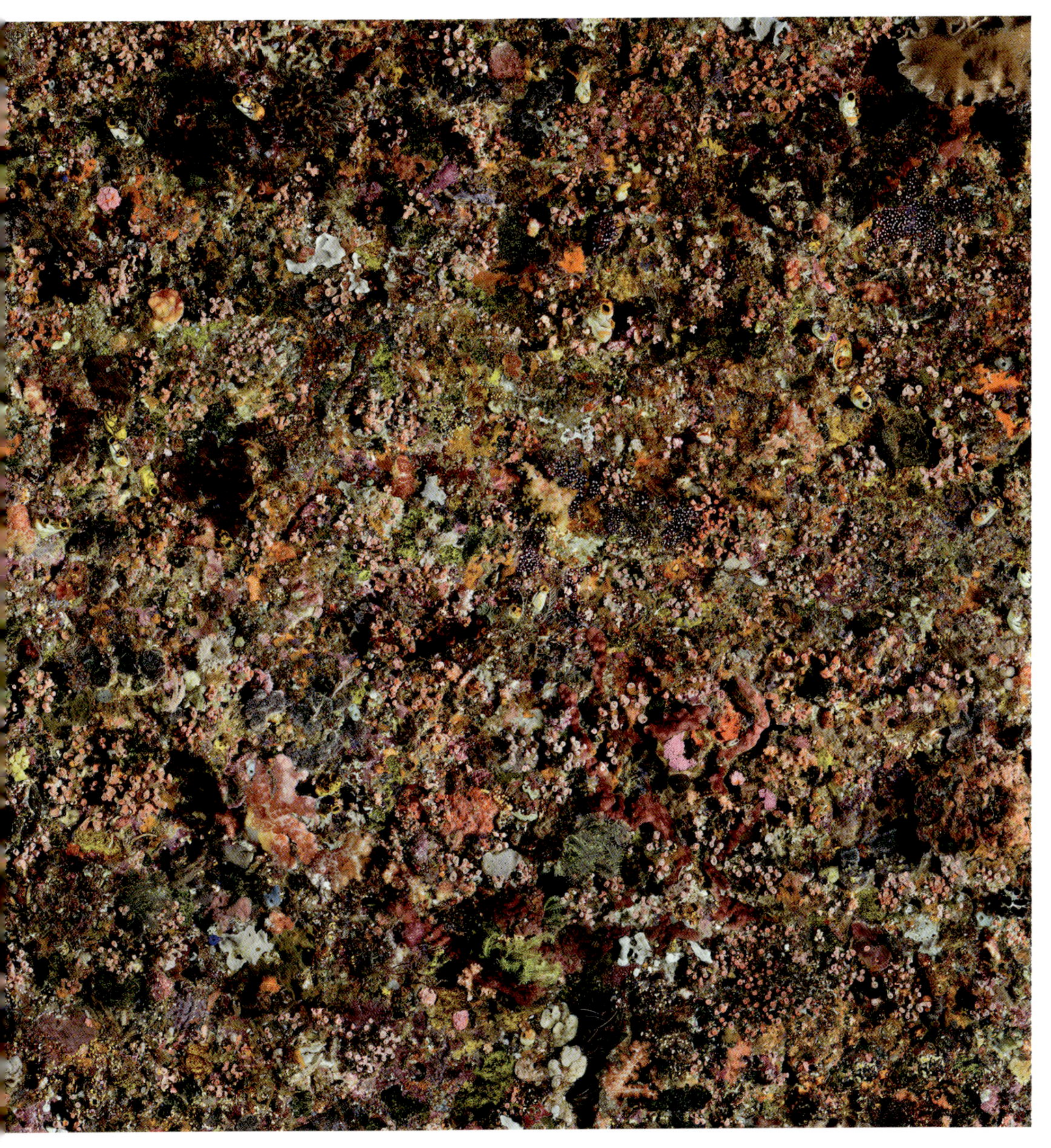

Edward Burtynsky, *Pengah Wall #1, Komodo National Park, Indonesia*, 2017–2018

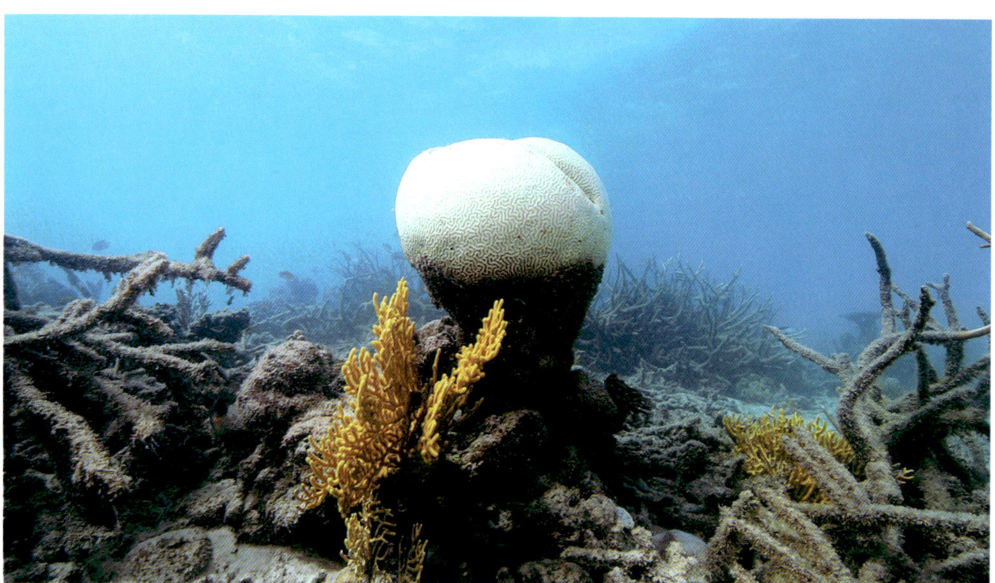

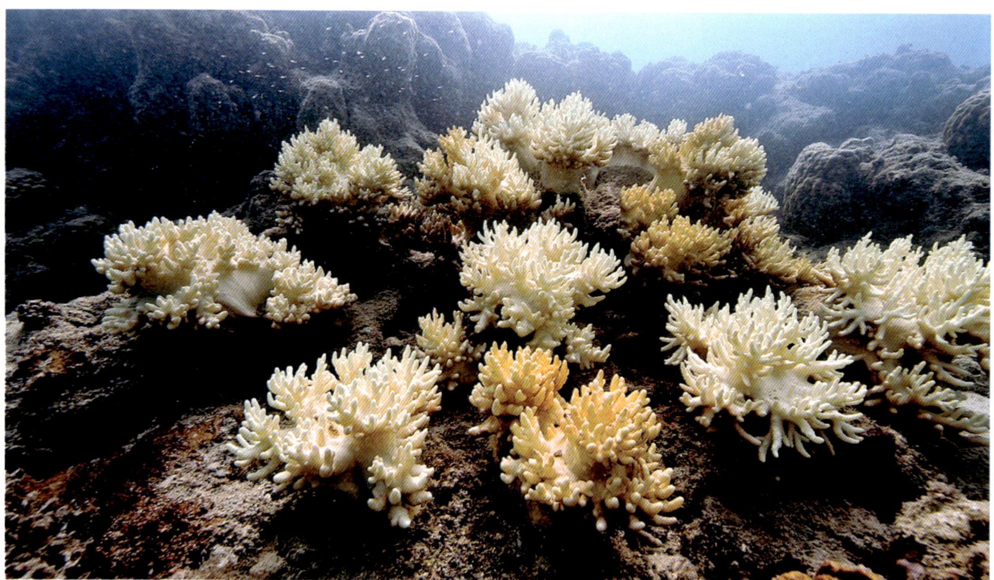

Jennifer Baichwal and Nicholas de Pencier
Bleached Coral, Great Barrier Reef, Australia, 2018

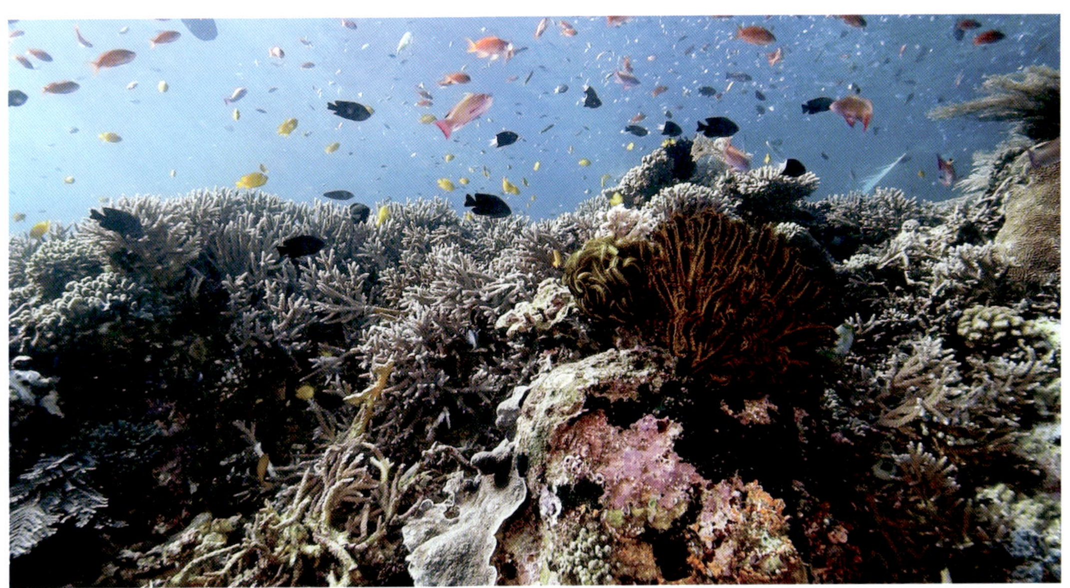

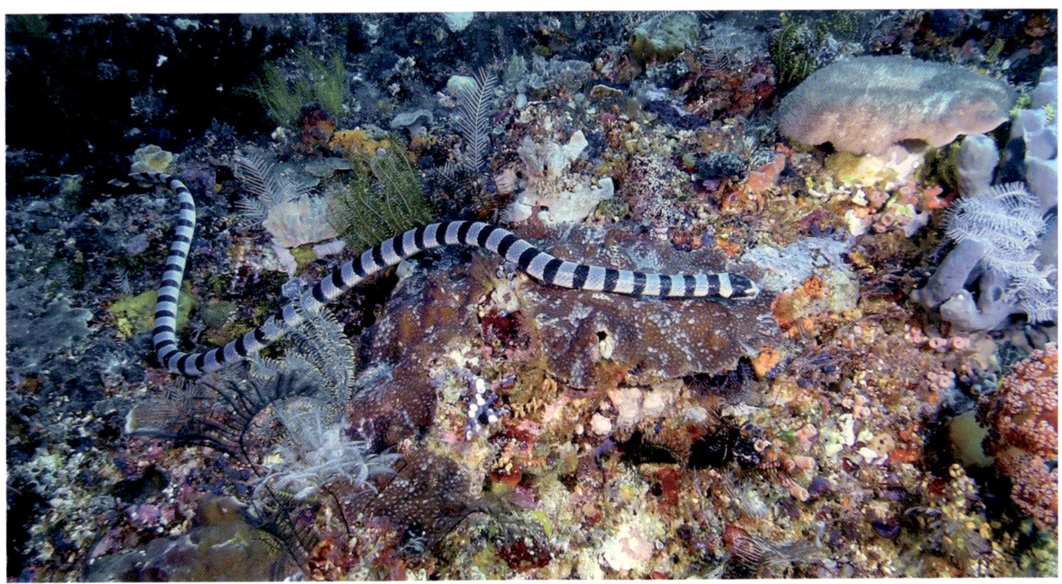

Jennifer Baichwal and Nicholas de Pencier
Panoply of Reef Life, Komodo National Park, Indonesia, 2018
Reef Portraits, Komodo National Park, Indonesia, and Great Barrier Reef, Australia, 2018

On March 20, 2018, the world learned that Sudan, the last remaining male northern white rhinoceros, had died. A grave loss for his long-time caretakers at Ol Pejeta Conservancy in Kenya, his death leaves only two members of his subspecies, Najin (his daughter) and Fatu (his granddaughter), and drives home the fact that all five of the world's diverse species of rhinoceros have been brought to the edge of extinction because of demand for their distinctive horns.[1] The horns, largely composed of the protein keratin (the same material as fingernails), have been prized for tens of centuries for their beautiful translucent colour when carved, and in traditional Chinese medicine for their supposed healing properties for fever, rheumatism, gout, and other disorders.[2] In the week following Sudan's death, Ol Pejeta Conservancy posted: "Fare thee well Sudan. You have done your work to highlight the plight of rhino species across the world; now the onus is on us to ensure that rhino populations thrive across our planet."[3]

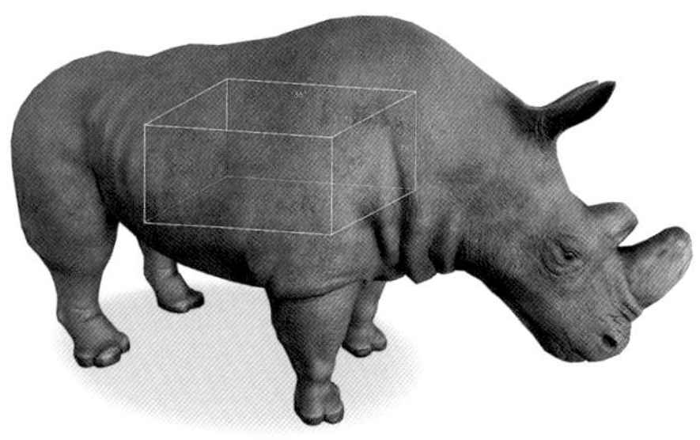

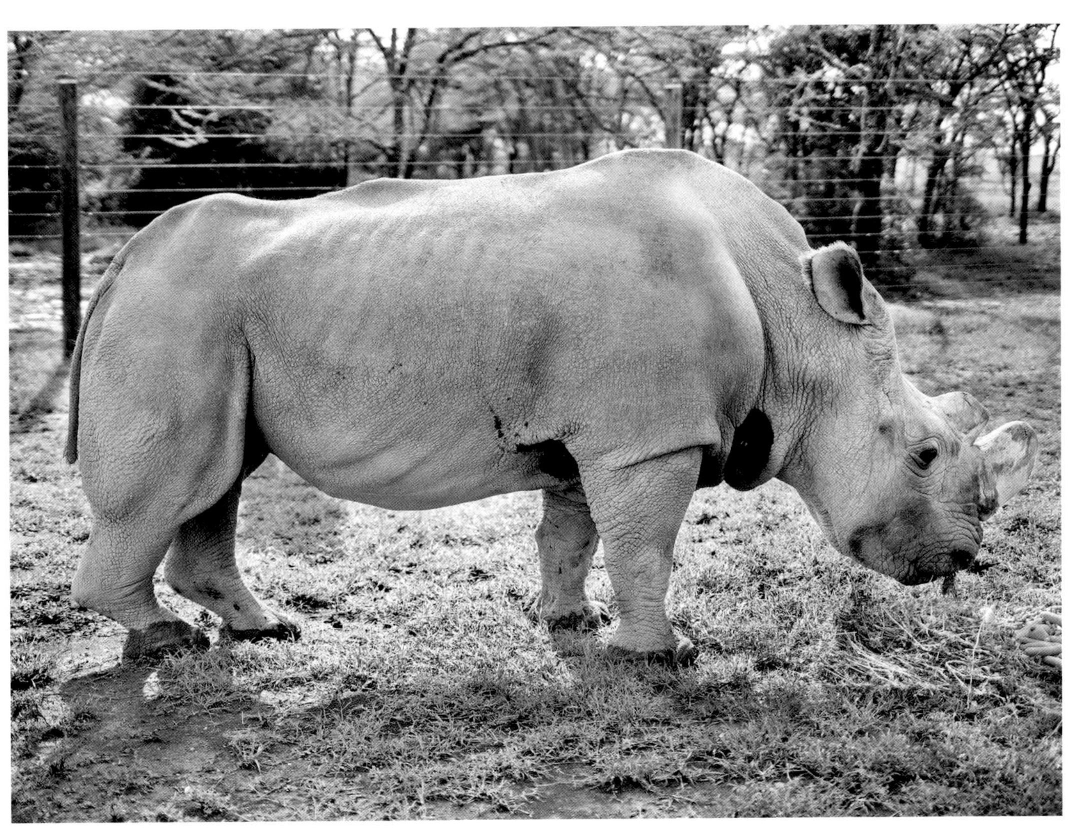

Edward Burtynsky, Jennifer Baichwal, and Nicholas de Pencier,
AR #4, Sudan, The Last Male Northern White Rhinoceros, Nanyuki, Kenya, 2016

Mass extinction is a marker of the Anthropocene, with many scientists agreeing we are currently in the midst of a sixth extinction—one resulting from human activity. Previous epochs have been delineated by abrupt ruptures in fossil records, such as the Paleocene-Eocene Thermal Maximum (PETM) or K-Pg boundary. Yet, while these mass extinctions took place over periods of millennia that might seem relatively brief (as most things are when measured against the entire geological time scale), the rapidity of the extinctions occurring at the hands of humans is by any measure extraordinary. A 2016 World Wildlife Fund report found that a staggering half of all animal species had seen a significant decline in population since 1970, with freshwater species most severely impacted.[1] It may be worth noting that when it comes to the public understanding of extinction, invertebrate surveys carry less emotional weight than those of mammals and majestic animals. Elephants, for example, are invested with potent meaning for human beings. Among the largest land animals on earth, they are significant in a great many cultures, representing everything from power and luck to wisdom and family bonds. As one of nature's gardeners, they are a keystone species, maintaining ecosystem balance and returning nutrients to the earth.

In recent years, illegal ivory poachers—heavily armed and employing military tactics—have killed tens of thousands of elephants across Africa. Their tusks are then smuggled through corrupt networks (often with the assistance of unscrupulous government officials) and sold to eager markets for use in upscale items: statuettes, jewellery, furnishings, and various trinkets.

It has been estimated that there were 167,000 elephants in Kenya in 1973.[2] The Great Elephant Census, published in 2016, found that number had dropped to 25,959.[3] People regularly die protecting the country's elephants from illegal poaching. In 1989, the Convention on International Trade in Endangered Species of Flora and Fauna (CITES) enacted a resolution commonly known as the "Ivory Ban."[4] That same year, the first-ever ivory burn took place in Kenya under then president Daniel arap Moi.[5] Effective as of 1990, all trade in elephant products was abolished, and ivory prices fell from $140 per pound to $5 per pound in one year.[6]

In 2016, the government of Kenya, under the leadership of incumbent president Uhuru Kenyatta, destroyed the largest-ever stockpile of ivory as a clarion call to halt all trade in ivory once and for all. One hundred and five tons of elephant ivory was burned,[7] representing between six and seven thousand elephants, as well as 1.35 tons of rhino horn.[8] The pyres were estimated to have a street value ranging between US$105 million and US$150 million.

Currently, demand for ivory is known to be concentrated in China, Thailand, Vietnam, the Philippines, and the United States. Yet there are encouraging signs. In March 2017, less than a year after the ivory burn in Nairobi National Park, Save the Elephants, a reputable wildlife group in Kenya, released a report indicating that the price of raw ivory in China had fallen by almost two-thirds in the last three years.[9] In 2015, both China and the United States pledged to shut down all ivory commerce. America's ivory ban went into effect in June 2016, and China's on December 31, 2017.[10]

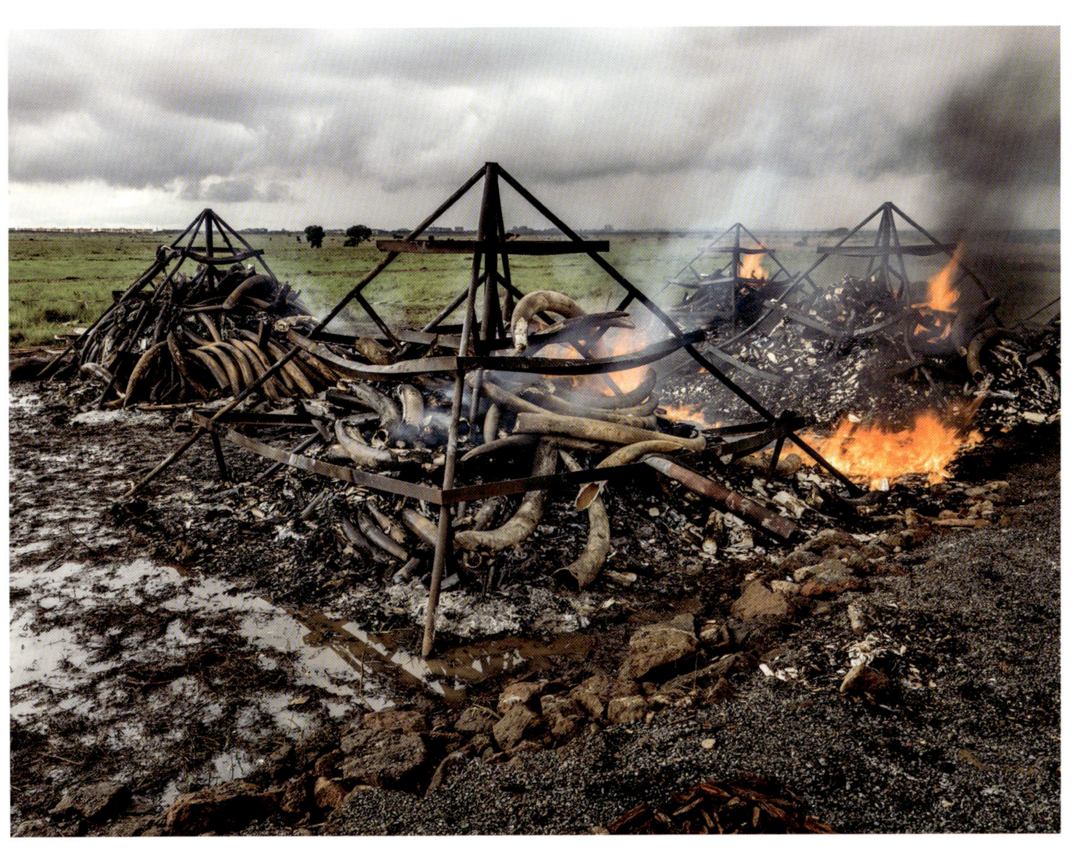

Edward Burtynsky, *Burned Ivory Tusks #1, May 1, Nairobi, Kenya,* 2016

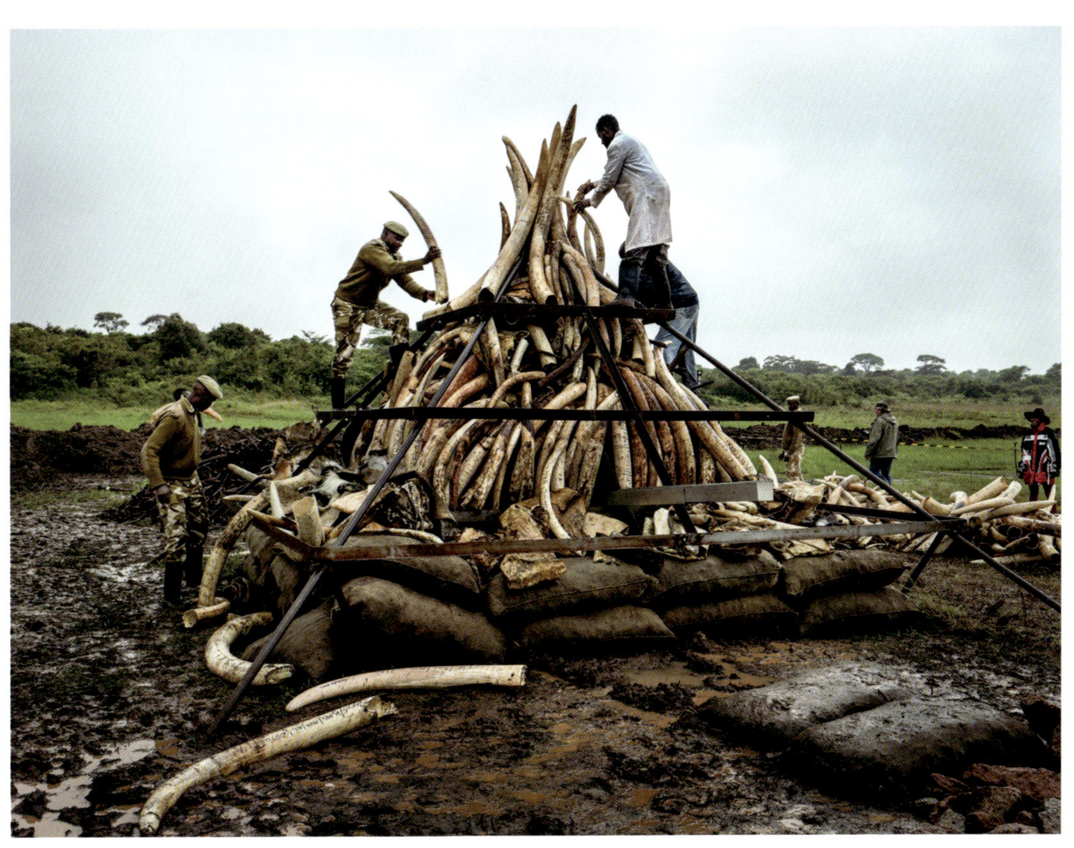

Edward Burtynsky, *Building Ivory Tusk Mound, April 25, Nairobi, Kenya,* 2016

On April 30, 2016, the largest ivory burn in history took place in Nairobi National Park in Kenya. For decades, the Kenyan government had been stockpiling elephant tusks and rhino horn that had been confiscated from poachers. They decided that a dramatic public incineration of this cache would make a bold statement to the world that there is no market for ivory. The day before the conflagration, the artists documented the largest pyramid of tusks, the so-called "President's Pile" that President Uhuru Kenyatta would set on fire the next day in a solemn ceremony in front of hundreds of dignitaries and press cameras. This photographic archive was achieved by taking more than 2,500 high-resolution stills of the three-metre-tall pile with a custom-built parallax DSLR rig from every angle possible. These images were then stitched together using specialized software into a highly detailed 3D mesh and texture map. The resulting virtual sculpture is a life-sized, photorealistic 3D model that can be explored using virtual or augmented reality platforms, and gives an exquisitely faithful representation of the complex monument of tusks that was burned to ash the next day. As a digital entity the sculpture is inherently ephemeral, and yet the striking verisimilitude of the virtual sculpture gives a visceral understanding of human-caused extinction.

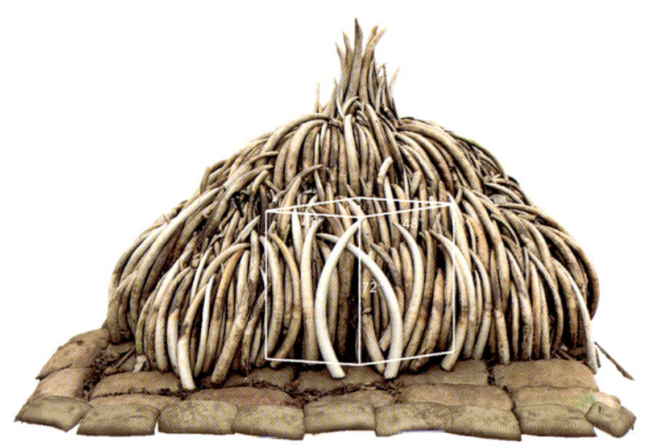

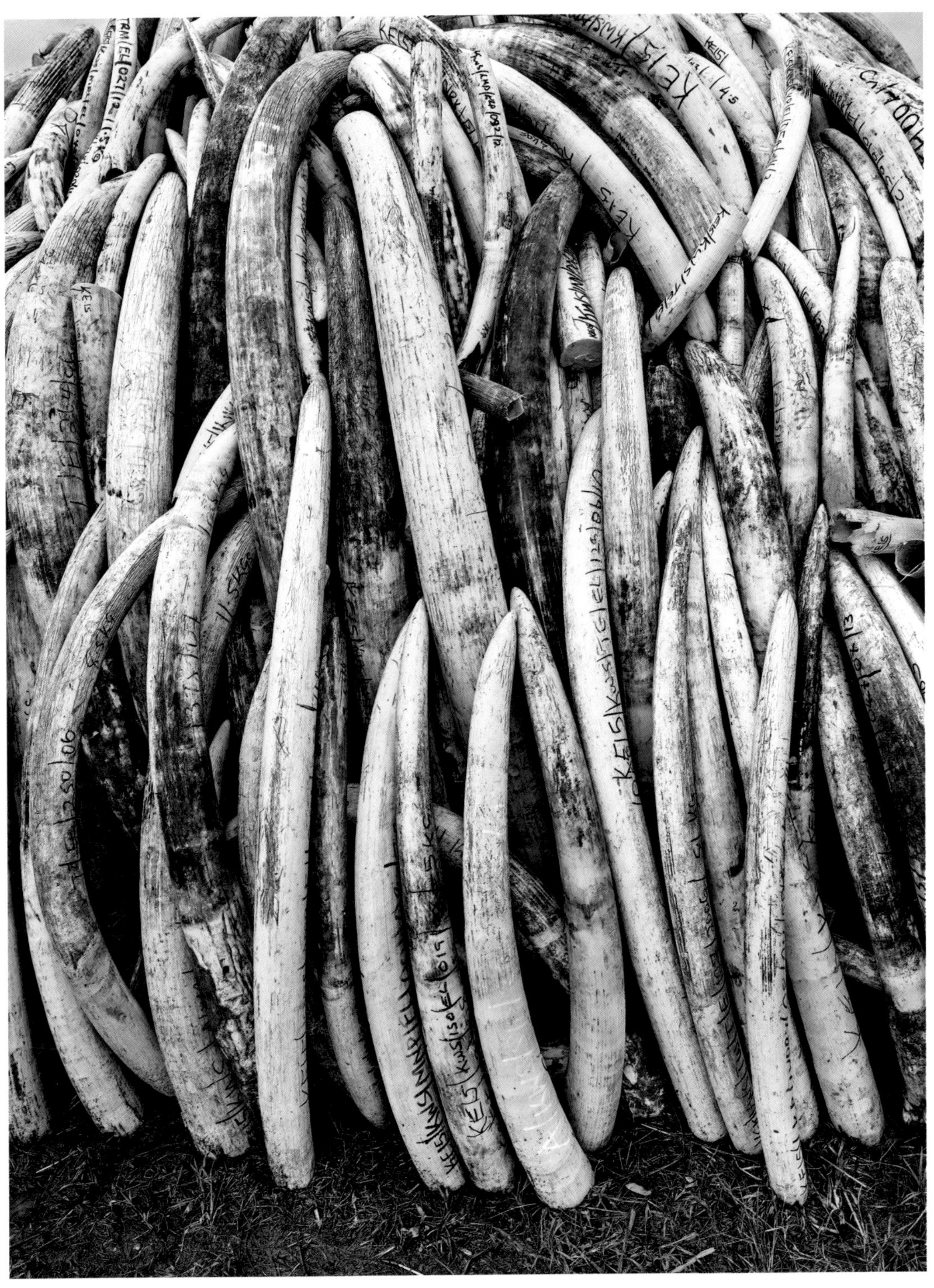

Edward Burtynsky, Jennifer Baichwal, and Nicholas de Pencier,
AR #2, President Kenyatta's Tusk Pile, April 28, Nairobi, Kenya, 2016

It is believed that, prior to European colonization, Africa may have been home to as many as 20 million elephants.[1] Today, approximately 352,000 remain in the continent.[2] In the lead-up to the historic tusk-burning event in 2016, Kenya Wildlife Service—in collaboration with a consortium of African and international anti-ivory groups, including Stop Ivory and the Tusk Trust—worked for over a year to organize and coordinate the burning of all of Kenya's confiscated stores of ivory. The tusk piles were built over the course of one week, during which the city of Nairobi witnessed a massive downpour. In the days prior to the event, the pyres required twenty-four-hour monitoring to prevent theft. Despite the torrential rain, the pyramid-shaped piles caught fire once they were lit. In an interview during the week of the burn, Dr. Winnie Kiiru, regional technical advisor with Stop Ivory, the organization responsible (along with Kenya Wildlife Service) for recording every tusk before it was burned, said:

> I am pleased that even if I didn't—or maybe many of us didn't—do what we needed to do to keep these elephants alive, no one is ever going to benefit further from their death. So burn on. I want to see every piece turn into ashes. And I'm going to be here until that happens. And then I will issue a certificate to KWS that says "You did it. We destroyed 105 tons of tusks. We destroyed 1.35 tons of rhino horn." May those animals rest. And no one is ever going to enjoy anything made from these tusks. These are ashes, and I'm happy.

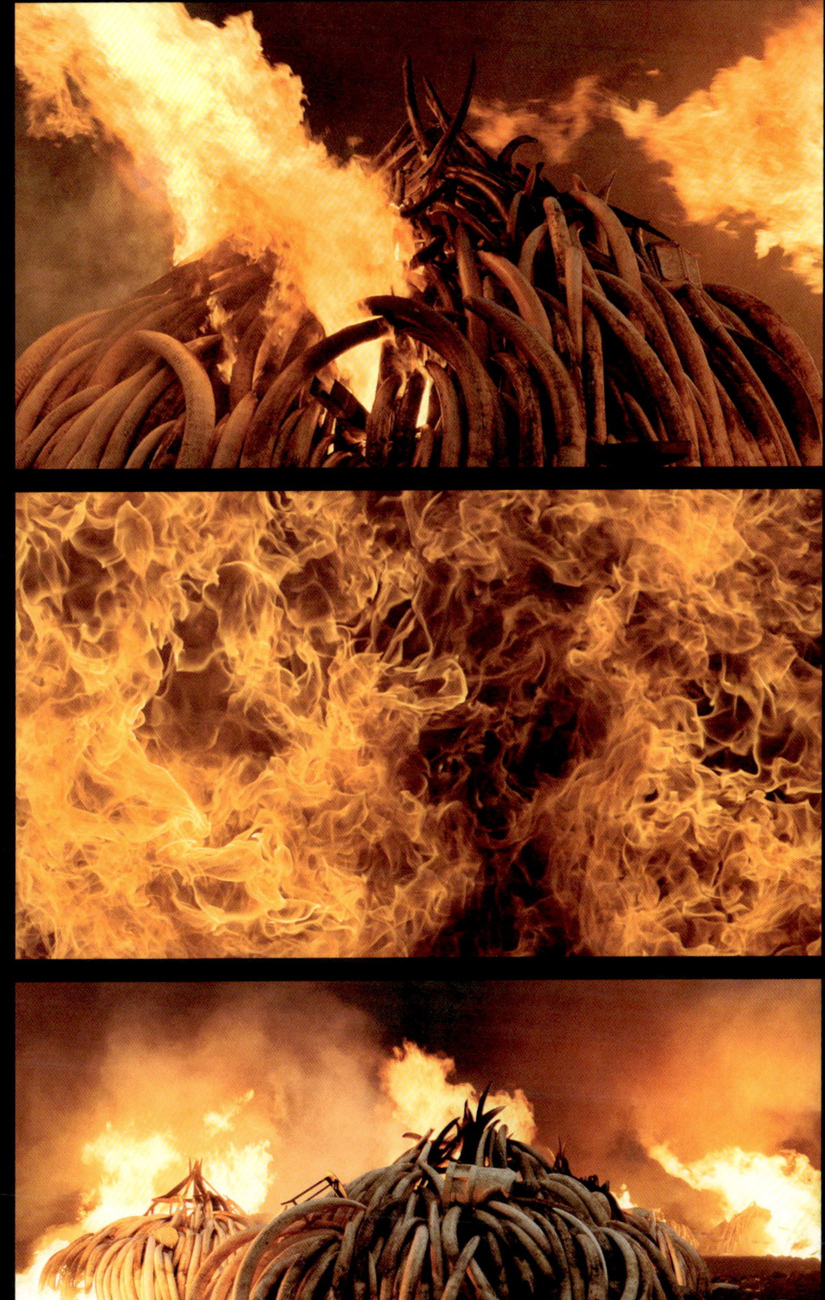

Jennifer Baichwal and Nicholas de Pencier, *Elephant Tusk Burn, Nairobi National Park, Kenya* (film stills), 2018

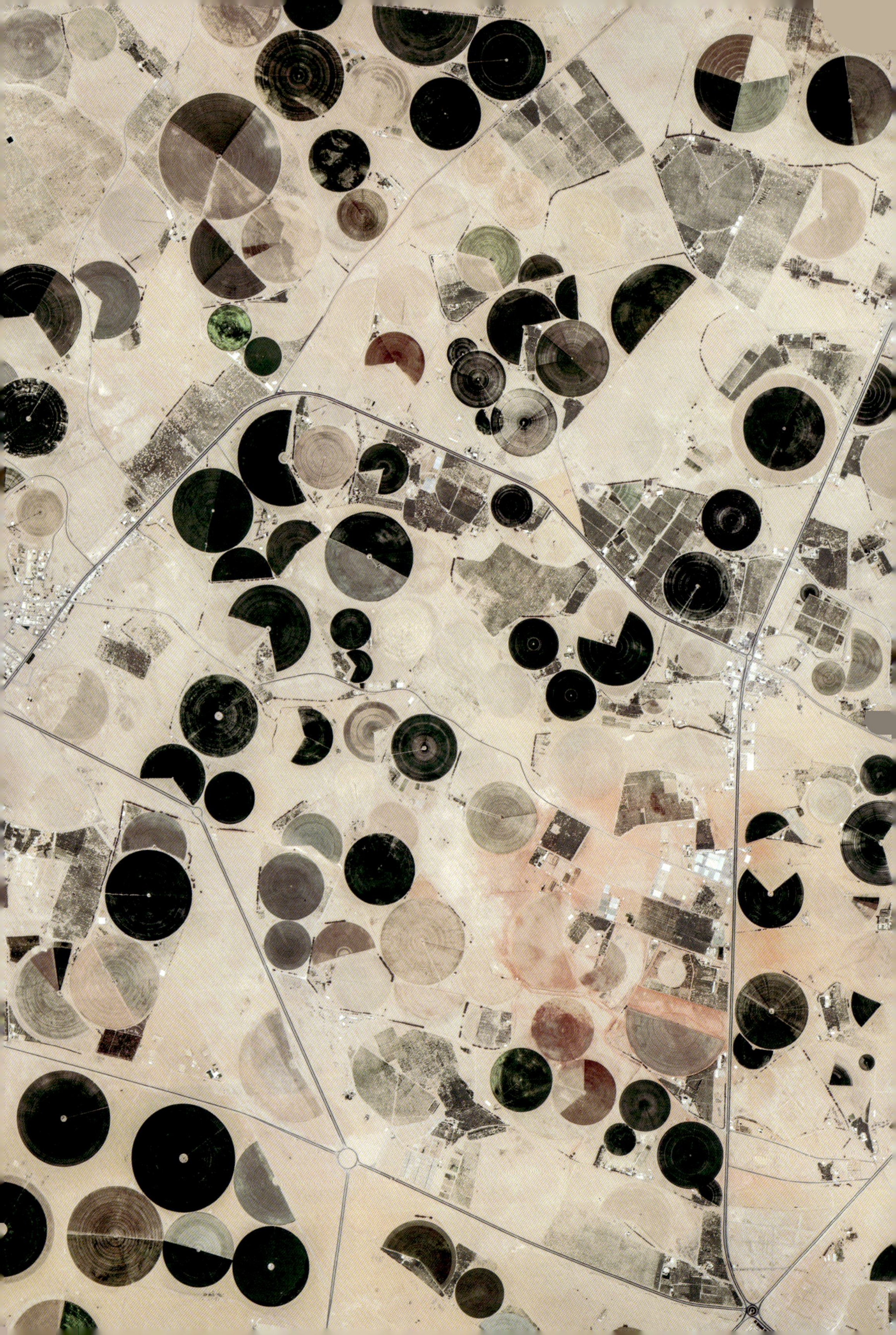

Life in the Anthropocene

Edward Burtynsky

My earliest understanding of deep time and our relationship to the geological history of the planet came from my passion for being in nature. As a teenager I loved to go on fishing trips, canoeing along the pristine isolated waterways of Ontario's Haliburton Highlands. That experience of wilderness left an enduring mark that still informs my response to landscape. I came to appreciate a state that exists without human intervention or disruption. I can remember learning, for example, that twelve thousand years earlier, in place of the lakes and trees where I cast my lures, a solid sheet of ice three kilometres thick had once covered the land. I realized that it was not only these wild places but the planet as a whole that is, and always will be, a dynamic, ever-changing system.

Our planet has borne witness to five great extinction events, and these have been prompted by a variety of causes: a colossal meteor impact, massive volcanic eruptions, and oceanic cyanobacteria activity that generated a deadly toxicity in the atmosphere. These were the naturally occurring phenomena governing life's ebb and flow. Now it is becoming clear that humankind, with its population explosion, industry, and technology, has in a very short period of time also become an agent of immense global change. Arguably, we are on the cusp of becoming (if we are not already) the perpetrators of a sixth major extinction event. Our planetary system is affected by a magnitude of force as powerful as any naturally occurring global catastrophe, but one caused solely by the activity of a single species: us.

OPPOSITE PAGE
Edward Burtynsky, *Satellite Capture, Near Buraydah, Saudi Arabia* (detail), 2018.

Between 1900 and 2000, the increase in world population was three times greater than during the entire previous history of humankind—a quadruple increase from 1.5 to 6.1 billion people in just one hundred years. Even in the brief span of my own lifetime, the earth's population has more than doubled to 7.2 billion. This phenomenal growth has coincided with a period of great scientific, cultural, and economic achievement, a deeper understanding of who we are, what we are capable of, and our relationship to the universe. However, as we access an ever-increasing body of knowledge with ever-increasing speed, we are also becoming more aware that this progress provokes as many pressing questions as it provides answers, with countless more people living in abject poverty and helplessness than those few with seemingly unassailable comfort, wealth, and power.

This extreme polarization has perhaps always been inevitable in the course of human evolution, the result of our species' innate social predisposition to greed. Unlike other species, there seems to be no end to our quest for food, comfort, shelter, sex—the fundamental necessities of survival that are now pursued in overdrive, far beyond our existential needs. But with this insatiable human striving has come our assault on the very planet that sustains us. During the last century, the use of global materials increased eightfold. Humanity currently uses almost 60 billion tons (gigatons) of material annually: biomass, fossil energy carriers, metal ores, industrial minerals, and construction minerals. It's clear that the expanding industrial metabolism is a major driver of global environmental change.[1]

I have come to think of my preoccupation with the Anthropocene—the indelible marks left by humankind on the geological face of our planet—as a conceptual extension of my first and most fundamental interests as a photographer. I have always been concerned with showing how we affect the earth in a big way. To this end, I seek out and photograph large-scale systems that leave lasting marks. At the heart of my challenge has been the pursuit of vantage points that best enable me to picture the relationship of these systems to the land. From my earliest shooting trips in the 1980s, I always sought to free my lens from ground level. In the early work on railcuts, homesteads, quarries, mines, and shipbreaking yards, I tried to find elevated spots to plant my tripod: a berm, a bridge, a rooftop or overpass; a perch that offered sweeping overviews of the subject matter. For my *China* and *Oil* projects I rented mobile scissor-lifts and bucket-lift trucks, raising myself up from the earthbound vantage points of my earlier work, taking in the wider spectacle.

The year 2006, however, marked a major change towards digital technology and a new way of generating work. I could now mount and electronically control my camera from a forty-foot pneumatic monopod. By using drones, airplanes, and helicopters, I could also achieve a bird's-eye perspective, rendering subjects such as transportation networks, mining, agriculture, and industrial infrastructure more expansively, capturing vistas that had eluded me until then. Rather than having my work be dictated by the limitations of topography or man-made structures, my lens could now literally fly.

In recent years, digital camera technology has become my main tool, allowing me to explore extended lens-based approaches. For *The Anthropocene Project,* large high-resolution murals measuring up to 3.6 by 7.3 metres are now made possible by capturing multiple frames and then seamlessly "stitching" them together, using software to create enormous printing files. Such has been the method behind several mural-size images I have recently made documenting extraordinary places: from the urban sprawl of Lagos and the Carrara marble quarries to the lush, primal forests of British Columbia and the coral reefs in Komodo National Park, Indonesia.

The internet has also become an invaluable tool, enabling my extensive location research but also allowing me to view and capture subjects from even greater distances. Using high-resolution satellite imaging, I have begun making compositions that extend my earlier interest in agriculture and its geometric interventions in the landscape, evident in my photographs of pivot irrigation sites from 2011. The image *Satellite Capture, Near Buraydah, Saudi Arabia* is the product of these new possibilities, gathering the visual data from a 160-square-kilometre sweep of industrial desert agriculture into one detailed image.

Extending the lens even further by using specialized 3D software, I am now able to capture and stitch together thousands of images of an object or a place, and render such scenes in full detail—as images that may be experienced in all their dimensionality. This has reignited my interest in looking at picture-making as a form of sculpture—something I experimented with, albeit more crudely, more than two decades ago. Dimensional imagery can now be experienced through virtual reality (VR) using headsets, and through augmented reality (AR)-enabled devices such as smartphones and tablets. I continue to be challenged by my practice, learning about technique, and also about the world around me. The work always leads me to new understandings, and new ways of perceiving.

Each place has its own story to tell, and its own distinctive impact. While producing the recent series on *Oil Bunkering* in the Niger Delta, for instance, I found myself confronted by some of the most foreboding landscapes I have ever seen. It's not since making my work on the Sudbury nickel tailings in 1996 that I had felt quite the chill of emotions that such an utterly devastated landscape can produce. Similarly, it was deeply humbling to witness President Uhuru Kenyatta's official burning of the largest pyre of ivory, just one of eleven such piles set ablaze during the historic Ivory Burn at Nairobi National Park in 2016. These funeral pyres, composed of elephant tusks, were immolated in order to prevent their entry into the market—an intervention that may change the future but cannot repair the damage of the past. In the cavernous potash mines beneath Berezniki, Russia, I was challenged by utter darkness, only to then experience the amazing psychedelic patterns and colours of the walls, not revealed until they were captured by our LED lighting techniques. (With my colleagues walking the mobile lights up and down the tunnels, literally painting the mineral-laden walls with light while I took the pictures, we were able to create myriad images that could later be joined together, compressing our collective time-based experience into static form.) This mine, one of five that I photographed in Siberia, comprises an estimated three thousand kilometres of tunnel.

By contrast, several of the photographs in *The Anthropocene Project* record some of the most verdant and transcendently beautiful places I have ever seen. Coming full circle, returning firmly to ground level and my reverence for the untouched wilderness I experienced as a teenager, I visited the remaining ancient redwood stands of British Columbia. It was important to me that this project also bring into sharper focus some of the wondrous ecosystems that are endangered, the beauty and biodiversity that we are at risk of losing.

This was particularly true of my first venture into underwater photography in 2016. In that year, I travelled with a support team to the Indonesian Island of Komodo, a UNESCO World Heritage Site, to dive and photograph some of the most pristine coral reefs still in existence. Using an underwater camera with high-powered flash units, and with the technical assistance of twelve divers, I spent several days mapping a specific section of coral wall, making several hundred high-resolution exposures, overlapping frames that could later be stitched together as a single continuous image. My ultimate goal was to create a finely detailed mural measuring ten by twenty feet—as close to actual scale as possible. Notably, this coral lies sixty feet beneath the ocean's surface. At that depth it appears as a dull, monochromatic blue. It wasn't until we began reviewing the shots that the full range and nuance of colour became breathtakingly apparent. The work at Komodo was perhaps the most challenging technical feat of my career, but it was also a headfirst plunge into one of the planet's most complex and dazzling ecosystems. That beauty spurs me on to ever more experimental ways of bearing witness.

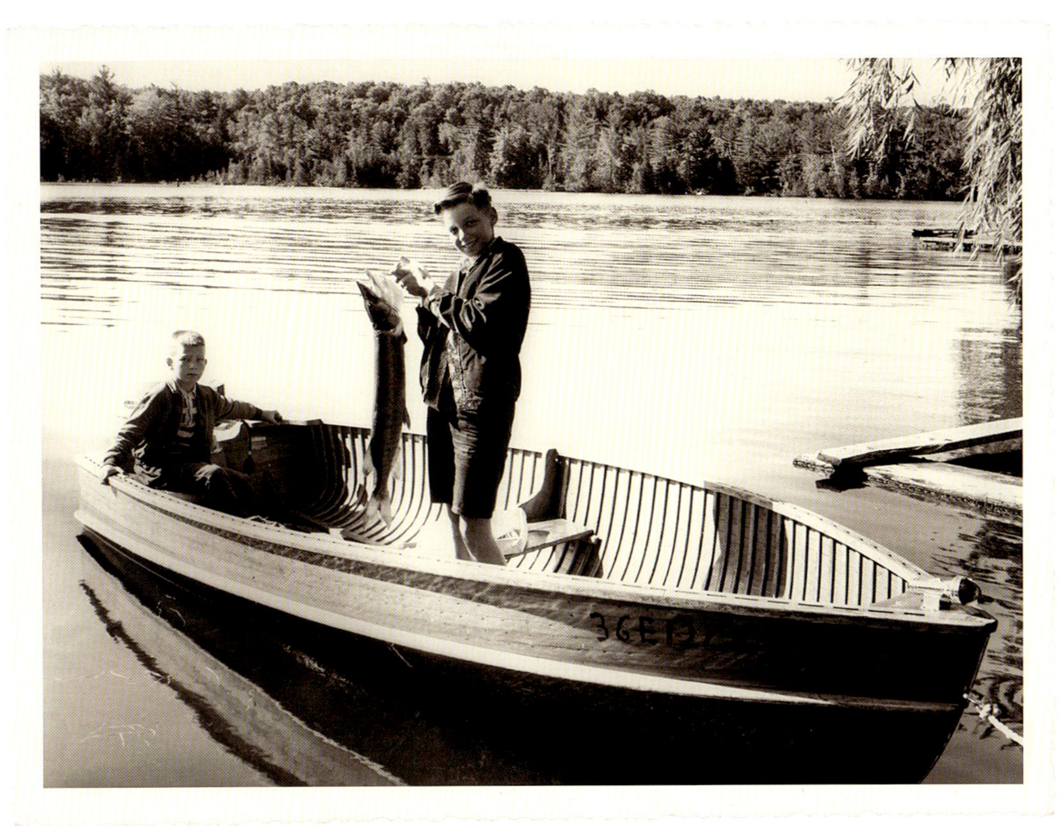

Eddy holding a Muskie with brother Michael
at the helm, 1968.

As a collaborative group, Jennifer, Nick, and I believe that an experiential, immersive engagement with our work can shift the consciousness of those who engage with it, helping to nurture a growing environmental debate. We hope to bring our audience to an awareness of the normally unseen result of civilization's cumulative impact upon the planet. This is what propels us to continue making the work. We feel that by describing the problem vividly, by being revelatory and not accusatory, we can help spur a broader conversation about viable solutions. We hope that, through our contribution, today's generation will be inspired to carry the momentum of this discussion forward, so that succeeding generations may continue to experience the wonder and magic of what life, and living on earth, has to offer.

Edward Burtynsky's imagery explores the human systems imposed upon our planet. His remarkable photographic depictions of large-scale industrial landscapes are included in the collections of over sixty major museums, including the National Gallery of Canada, the Museum of Modern Art and the Guggenheim Museum in New York, the Reina Sofía Museum in Madrid, and the Los Angeles County Museum of Art. Burtynsky's distinctions include the TED Prize, the Outreach Award at the Rencontres d'Arles, the Roloff Beny Book Award, and the Rogers Best Canadian Film Award. He sits on the board of directors for the Scotiabank CONTACT Photography Festival and the Ryerson Gallery and Research Centre, and is co-founder of the Scotiabank Photography Award. In 2006, Burtynsky was awarded the title of Officer of the Order of Canada and in 2016 he received the Governor General's Award in Visual and Media Arts. Burtynsky currently holds eight honorary doctoral degrees. His books with Steidl are *China* (2005), *Quarries* (2007), *Oil* (2009), *Water* (2013), and *Salt Pans* (2016).

1. Krausmann Fridolin, Simone Gingrich, Nina Eisenmenger, Karl-Heinz Erb, Helmut Haberl, and Marina Fischer-Kowalski, "Growth in Global Materials Use, GDP and Population During the 20th Century," *Ecological Economics* 68, no. 10 (2009): 2696–2705.

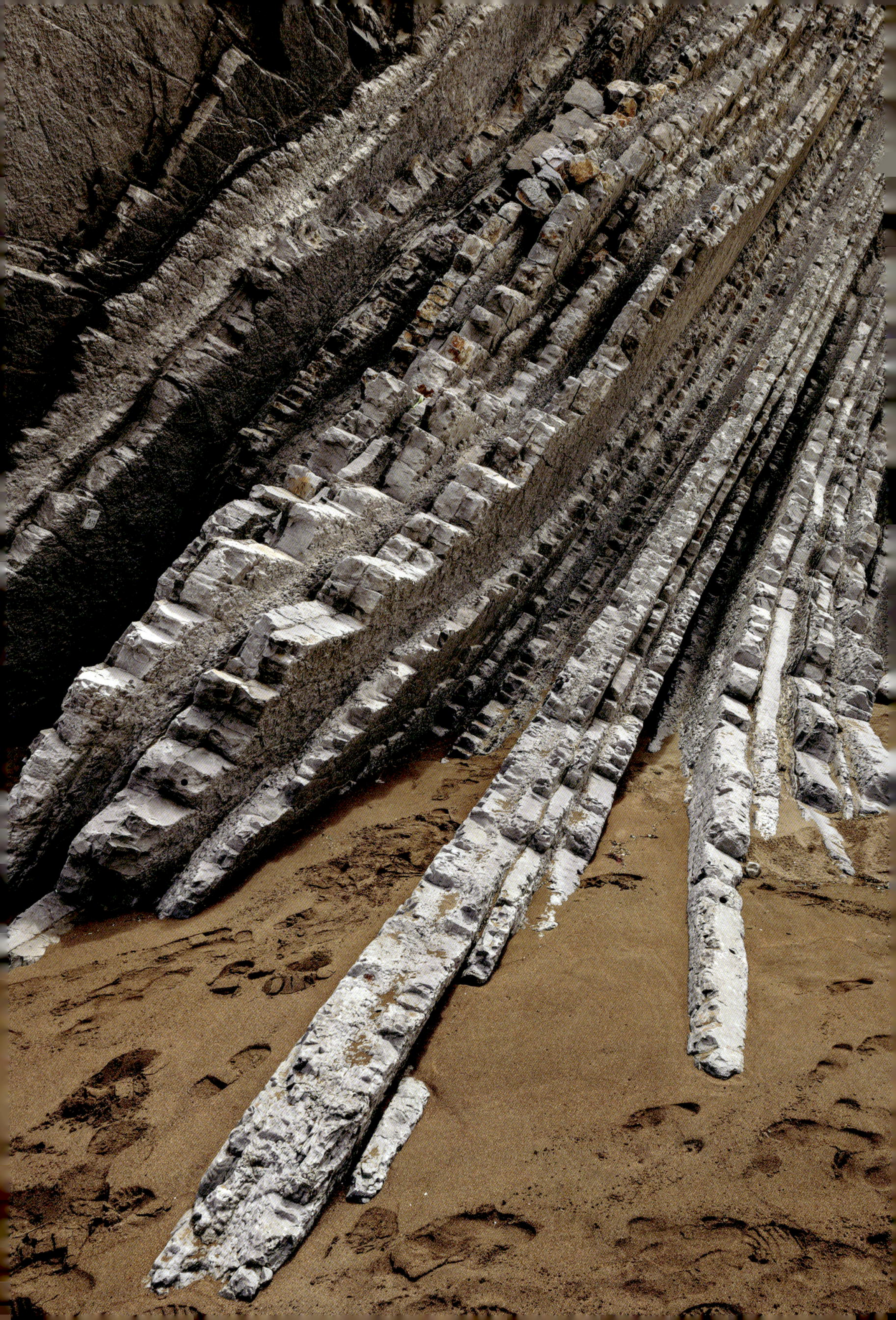

Our Embedded Signal

Jennifer Baichwal

To think in geological time is almost impossible.

Our first significant shoot for *The Anthropocene Project* was in Zumaia, on the Atlantic, in the Basque region of Spain. Itzurun Beach has some of the most dramatic, stunning cliffs in the world—"flysch"—formed over millennia from the ocean floor tilting up to vertical through tectonic activity. The resulting coastline represents 60 million years of earth's history. Sixty million years. How to even comprehend this breadth of time? How to comprehend the brevity of our species' place within it?

I, along with my husband and collaborator Nicholas de Pencier, have been making various kinds of films, mainly documentaries, for twenty-five years. These have always started by grappling with the ethics of engagement, and have always tried to extend documentary form into something more existentially and philosophically weighted, beyond traditional text/visual relationships. We have now worked with Edward Burtynsky on and off for thirteen years, three films, and three bodies of work. *The Anthropocene Project* is a culmination of all of the conversations we have ever had about art's capacity to provoke change, as well as the merits and drawbacks of doing this experientially: to not preach, harangue, or blame, but to witness, and in that witnessing, try to shift consciousness.

There is some danger to this approach because you can argue that the scale and rapidity of planetary destruction requires something more urgent, something that more stridently signals emergency. Point taken. But I still believe that lateral exploration, the open-ended conversation, can provoke transformation more deeply than hard argument.

When we made *Manufactured Landscapes,* the 2006 film about Burtynsky's photographic essay of the industrial revolution in China, I knew from the beginning that I didn't want to make a traditional artist portrait; I wanted to try to extend the meaning of his photographs into film, to intelligently translate one medium into another. This would prove to be harder than expected.

OPPOSITE PAGE
Edward Burtynsky, *Basque Coast #2, UNESCO Geopark, Zumaia, Spain* (detail), 2015.

One challenge was to express scale in time. At one point, in Xiamen, we filmed a factory floor the size of eight football fields. After we had walked the huge place for a few hours, trying to figure out how to best convey it, we saw a golf cart gliding past the endless rows of people working. My cinematographer and collaborator Peter Mettler said: "Let's do it as a dolly." The nine-minute continuous shot that resulted was a perfect articulation of scale in time. I knew, after we had blown through a week's worth of 16mm film stock in a day, that it would be the opening scene. Eleven years later, for this project, Nick and I strapped a RED Epic camera to the front of a train which then travelled the longest railway tunnel on earth: fifty-seven kilometres through the Swiss Alps. It takes about 20 minutes. Approximately 28 million tonnes of material was excavated to make this tunnel. Scale in time.

Another problem was how to honour particularity in context, to find the right ethic and dialectic between scale and detail. How can you go all over the world to places you are not intrinsically connected to and hope to reveal anything significant or true? How do you proceed without arrogance and with as much interrogation of your own bias as possible? When I use the word "true" I do not mean objective, dispassionate reportage, which we all know doesn't exist. I mean experiential knowledge that opens up empathy and understanding.

In Burtynsky's scale views, the vicissitudes and struggles of daily life are implicit, not explicit. It was crucial to me that we express them in some way; that we engage with context in such a way that it might reveal itself. It is why we spent so long with the women making the spray mechanisms for irons in the vast Xiamen factory. We were not allowed to interact, so we had to find another way to be with them. We focused on their hands and faces as they did their repetitive labour that continued day after day, month after month, year after year. We could witness and recognize, up close, the painstaking effort that goes into making things that many of us use and discard unconsciously.

Burtynsky's photographs take people to places they are connected to or responsible for but would never normally see, and he finds the wide view that communicates them in one frame. Having been with him in these landscapes, I know that isn't easy. Our films—the long documentaries and the short installations—try to explore and extend the narratives of place by deepening understanding of context and detail, then reflecting this back onto ourselves.

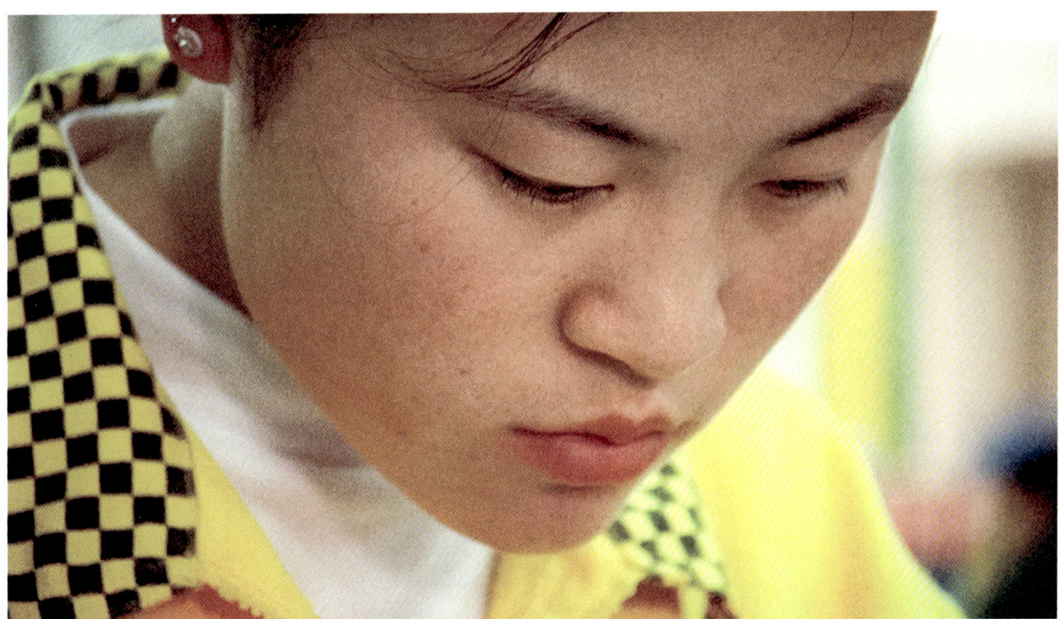

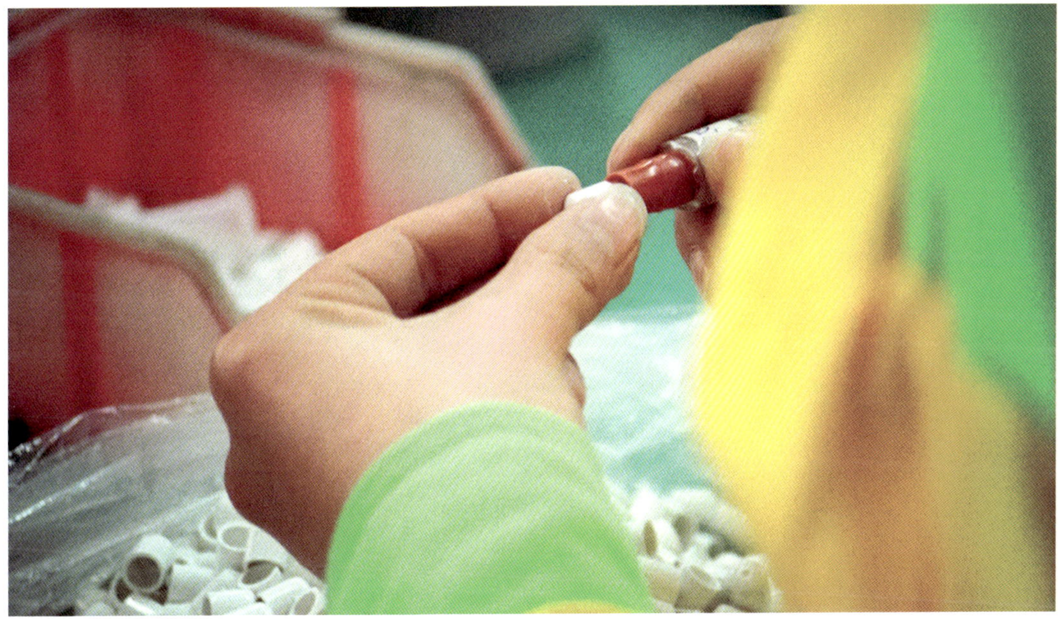

TOP
Woman concentrating on the task at hand, Xiamen, China, *Manufactured Landscapes* (film still), 2006.

BOTTOM
Assembling spray mechanisms for irons by hand, Xiamen, China, *Manufactured Landscapes* (film still), 2006.

In *Watermark*, for example, we explored the Buriganga River, which is the main waterway of Dhaka, and crucial to the city's daily function. The Hazaribagh leather tanneries in the city centre use copious quantities of water from the river and also discharge their toxic effluent directly into it, nearly 20 million litres every day.[1] There is no sustained enforced regulation on this. This was a point in the film where we could have gone to a scientist or water expert to describe the problem in detail, and what needed to be done to address it. Instead, we see a woman in a sari standing at waist level in the river; she holds her nose and then submerges herself. We see a father on the bank scoop the river water up to wash his son's hands and face. The son has the stoic, trusting look of a child waiting for an adult to finish administering some boring but necessary daily ritual. The father is being a father: bathing his child, as any parent has done thousands of times. He does not know that the effluent in the river makes the water toxic and carcinogenic.

The leather treated in the Bangladesh tanneries and the shoes, purses, and belts made close by are sold all over the world. Hazaribagh is a place we are responsible for but would never normally see.

For *The Anthropocene Project*, we visited the industrial city of Norilsk, two hundred miles north of the Arctic Circle, which was founded as a slave labour camp in 1935. It has the world's largest coloured metal mine (copper, nickel, palladium) and the largest heavy metals smelting complex in the world. These metals are used in everything from cell phones to water pipes, car radiators and catalytic converters to dishwashers.

Here we were detained and fingerprinted by the Federal Security Service for being "journalists who came into the city under false pretenses," because we approached three women to ask them about their lives. The women were crane operators in a copper smelter, and on their coffee break. Here is some of what they said:

> "We work with the slingers. They give us signals and we need to always be attentive and careful. We are responsible for the lives of other people. Many people love Norilsk, but some want to leave. Greenery is severely lacking here. Oxygen is too, as a result. We have the polar day, and the polar night. Those have their positives and negatives too. Most of those who want to leave do, but then they miss Norilsk. It's not an ordinary city. At the beginning, it takes some getting used to, but when you adjust, it pulls you in, and becomes your own. You become a romantic. You see beauty in a flower that's bursting through the stone."

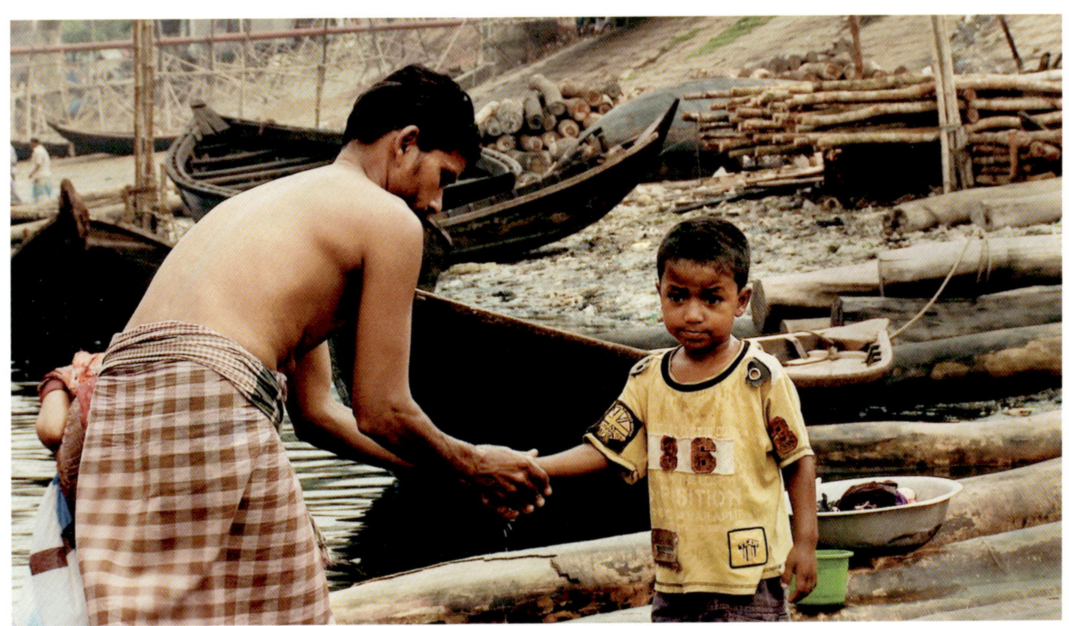

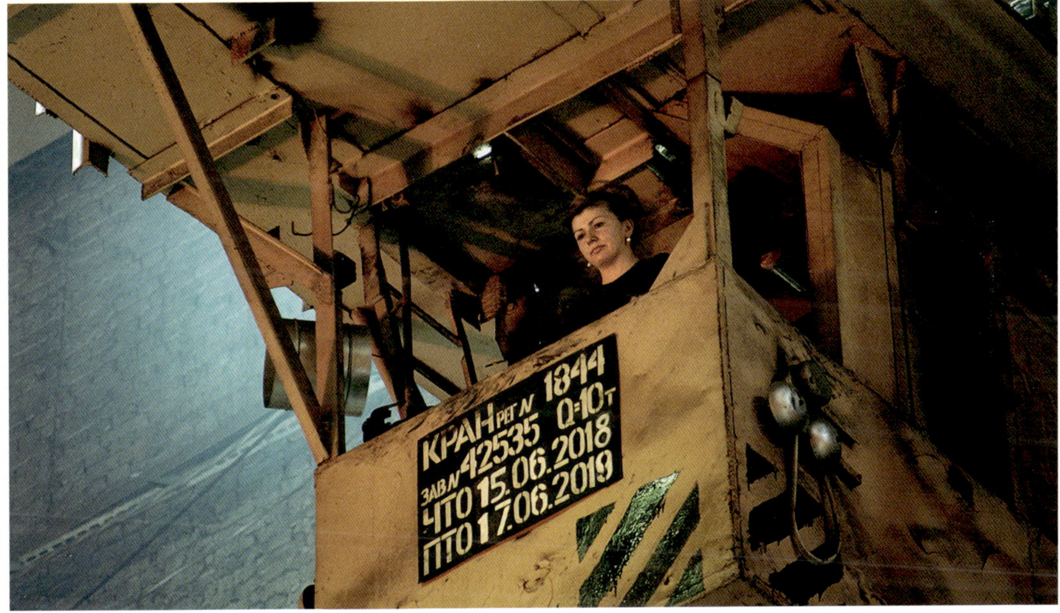

TOP
Father bathes son in Buriganga River, Dhaka, Bangladesh, *Watermark* (film still), 2013.

BOTTOM
Woman operating crane at smelter, Norilsk, Russia, *Anthropocene* (film still), 2018.

Norilsk is a closed city, meaning that you need special permission, even as a Russian, to visit. It has also been named by Pure Earth (formerly the Blacksmith Institute) as one of the most polluted places on earth. Five hundred tons each of copper and nickel oxides, and 2 million tons of sulphur dioxide, unscrubbed, are released into the air every year. The life expectancy of factory workers in Norilsk is ten years below the national average.[2]

We are now at the point where humans—and let's face it, primarily we humans who live in developed countries and have historically colonized, polluted, taken from, and destroyed our way around the globe—are changing the planet and its systems more than all other natural forces combined. That's quite a feat for a species that has only been around for about 200,000 of the earth's 4.5 billion years. Which brings me back to Zumaia, and the original question. How to even comprehend this breadth of time? How to convey, despite our brevity as a species, the magnitude of our impact? Anthropocene in a scientific and geological sense means that we are now everywhere, all the time, and even in the rocks—those dense, mysterious receptacles of the planet's history. Our signal is embedded in the strata. For this project, we extended our respective practices beyond film and photography to film installations, murals, photogrammetry, virtual reality, and augmented reality; what Burtynsky calls "photography 3.0." We then went all over the world seeking and engaging places, people, and contexts that would experientially shift consciousness about what it looks like for humans to be everywhere, affecting everything, all the time. With some exceptions, it doesn't look great. But a shift in consciousness can be the beginning of change.

Jennifer Baichwal has directed and produced documentaries for over twenty years. Her films have screened and won awards nationally and internationally. *Manufactured Landscapes,* about the work of Edward Burtynsky in China, was released in twelve countries and received numerous distinctions, including Al Gore's Reel Current Award and Best Canadian Feature Film at TIFF 2006. *Watermark* (co-directed by Edward Burtynsky, produced and filmed by Nicholas de Pencier) premiered at TIFF 2013 and was a special presentation at the Berlinale in 2014. It was subsequently released in eleven countries and won the Toronto Film Critic's Association prize for Best Canadian Film 2014 and the Canadian Media Awards prize for Best Documentary in the same year. *Anthropocene* is Baichwal's tenth feature documentary.

1. H. L. Paul, A. Paula M. Antunes, Anthony D. Covington, Paul Evans, and Paul S. Phillips," Bangladeshi Leather Industry: An Overview of Recent Sustainable Developments," *Journal of the Society of Leather Technologists and Chemists* 97, no. 1 (2013): 25–32.

2. "Norilsk, Russia," *Top Ten Most Polluted Places,* Pure Earth, 2007, accessed April 23, 2018, http://www.worstpolluted.org/projects_reports/display/43.

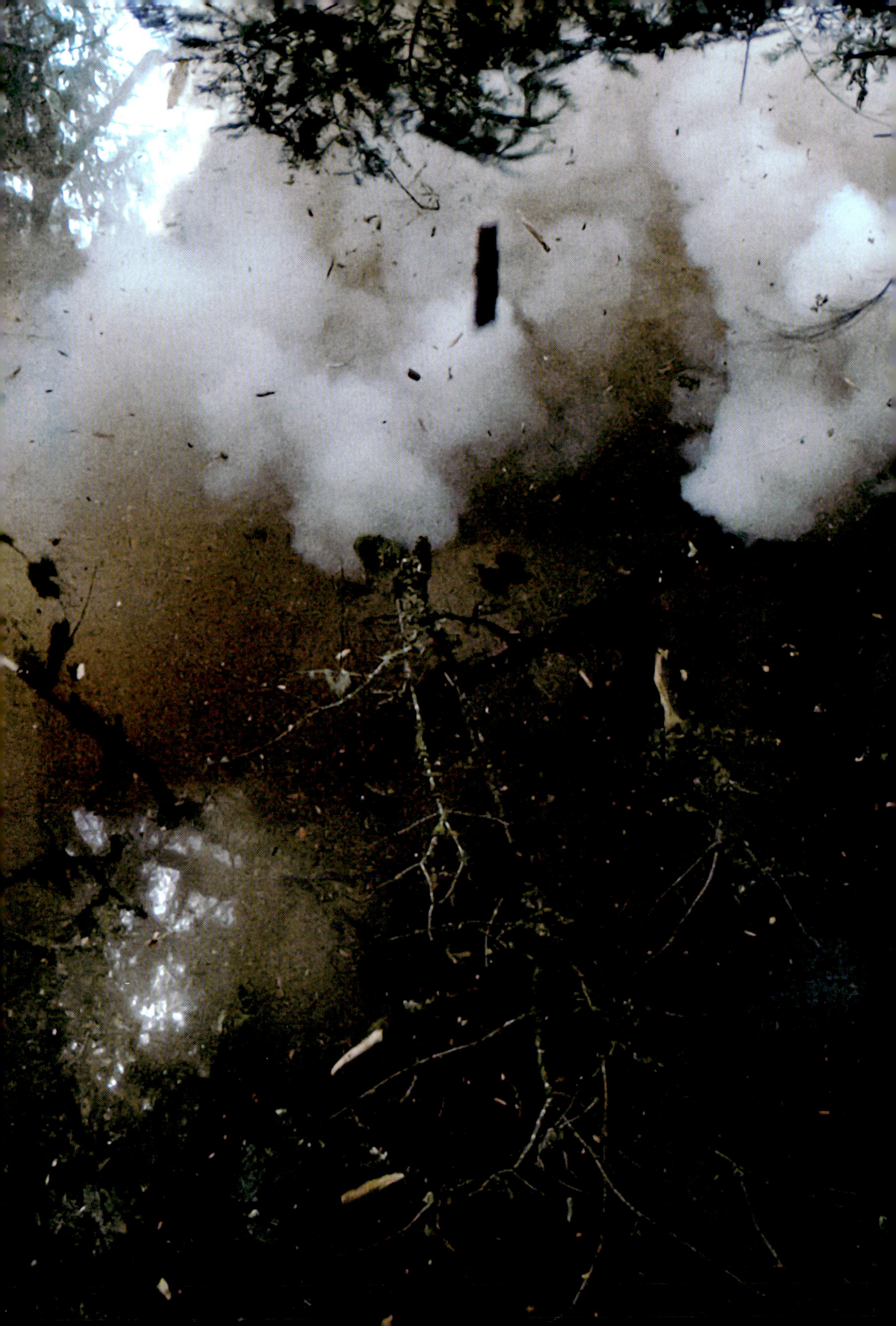

Evidence

Nicholas de Pencier

My grandparents would have had difficulty understanding the concept of the Anthropocene. Ironically, theirs was the generation that presided over the "Great Acceleration" after the Second World War, which the scientists of the Anthropocene Working Group are touting as the definitive start to the human epoch. Anthropogenic transformation of the earth has appeared incrementally throughout human history. Land was cleared for agriculture, various animals were hunted to extinction even as invasive flora and fauna were dispersed through human migration, and of course the burning of fossil fuels allowed an exponential increase in all human activity. My grandparents would have been conscious of those dynamics. They sang hymns in church decrying "dark satanic mills" and derided lack of intelligence by calling people "dodo birds." But for them, human technological progress was inherently positive. It was a natural extension of the innate impulse to expand to fill the carrying capacity of the environment around you.

This impulse had remained unchanged since the beginning of genetic memory. In Canada in the first half of the last century, the natural world still had an aura of monumentality. How could my grandparents imagine that the burgeoning technology that surrounded them—their newfangled aluminum cooking pots, ubiquitous plastics, or miracle chemicals like DDT (advertised as so safe you could drink it)—would represent a tipping point for humans versus the earth's systems? Not that they were wasteful or profligate with the things that progress afforded them. Quite the opposite. They were pathologically frugal and abhorred waste of any kind. They saved every elastic band that ever came through their mail slot. But for them, nature still had dominion over all living things and it would be vain to suggest otherwise. Note that there were only 1.6 billion people on earth when they were born. The evidence of the scale of the impact of human activity was not available to them in the way it is to us now.

OPPOSITE PAGE
Jennifer Baichwal and Nicholas de Pencier, *Exploding Danger Trees #2, Cathedral Grove, British Columbia, Canada* (film still detail), 2018.

Presenting our take on that "evidence" is at the heart of *The Anthropocene Project*. As someone who is compelled to chronicle places and events with cameras, I am drawn to those subjects which inspire me and which worry me. The iconic themes and locations we have chosen for *Anthropocene* do both. It's hard not to marvel at the engineering ingenuity of the massive industrial sites we filmed, and equally hard to ignore the devastation they represent. The Anthropocene is rife with staggering human folly. Many of the environments we visited were so toxic they elicited a disorienting revulsion and palpable relief when we left. Internalizing the power of that reaction predisposes me to believe all the scientists who are loudly ringing alarm bells of dire consequences in so many sectors on our planet. But there is also much that is positive in the places we have captured in this project: inspiringly smart people dedicating themselves to solving our planet's problems. There is a danger that when you take on the task of chronicling the hubris of the Anthropocene, you will come off as trying to seem separate from, and even superior to, those who seem to be responsible. It's important for me that we don't point fingers or disavow our own culpability. I live in the real world and need the same practical solutions that everyone else does. It is therefore my responsibility to use my camera as a mirror, not a hammer: to invite viewers to witness these places and react in their own individual fashion.

Cameras are a natural evolution of our ability to abstract from reality. The power and utility of being able to capture a moment in time and space, examine it, and share it with others is addictive. This uniquely human capacity for recording abstractions is ultimately the reason the Anthropocene can exist at all. Drawings, stories, texts, maps, blueprints, models, diagrams, instructions: human knowledge is cumulative and we stand on the shoulders of those who came before us. By the time photography appeared, many other forms of expression and dissemination had played their part in our achievements and our meta-consciousness. But photographic representation of reality was different. It was seemingly less subjective: less mediated by brushstrokes, or wordsmithing, or other editorial impositions. Its physical, chemical nature made it feel closer to an idea of truth. A dispassionate scientific credibility anchored the aesthetic. The photons were there already—the photographer needed only to catch them with the camera, and then stand back and marvel at the results. Somehow that perceived objectivity felt more real, and so conferred power and weight to the medium. It still does, and somewhere in that complex semiotic soup lies the answer to why lens-based media are the most popular ones of our age. They represent the most technological, most anthropogenic, of the art forms, and that is why we feel confident deploying them for this project.

The Anthropocene *is* full of burning questions. Does it exist? If so, how do we define it? When did it start? How might it end? We are drawn to these questions because they are a way of framing the biggest practical questions of our age, and also our undying and unanswerable philosophical dilemmas. Is the Anthropocene good or bad? Who are the heroes of the Anthropocene? The villains? The images that make up *The Anthropocene Project* are meant to ask these questions and be a starting point for a journey, not a resolution.

I wonder if my children will have difficulty comprehending the idea of the Anthropocene. Marshall McLuhan said that if you want to know about water, don't ask a fish. My kids are so deeply immersed in the technosphere that imagining a time when our web of digital communication did not exist will be a challenge for them. This is another ambition of *The Anthropocene Project:* to invite the viewer to step out of the bubble of frenetic modern existence and contemplate human activity on a planetary scale and in geological time. When I stand barefoot on the 3.9-billion-year-old granite of the Canadian Shield, I get a visceral feeling that I am what geologists call a "fleshy transient." In the earth's history, my little moment of being is mightily insignificant, except that I am conscious of it. Can we extrapolate from that feeling to an understanding of the Anthropocene? Can we put our species' momentary blip in time and space in a greater context? And if so, can we harness that mindfulness to affect our future? In this project we are following the scientists' rationale to define the start of the Anthropocene, but how the story ends is still very much up for grabs. My hope is these images we have worked so hard to collect will have some pull with those who experience them, such that they are drawn-in and feel implicated in a transformational way.

Nicholas de Pencier is a documentary director, producer, and director of photography. Selected credits include *Let It Come Down: The Life of Paul Bowles* (International Emmy), *The Holier It Gets* (Best Canadian Doc, Hot Docs), *The True Meaning of Pictures* (Gemini, Best Arts), *Hockey Nomad* (Gemini, Best Sports), *Manufactured Landscapes* (TIFF Best Canadian Feature; Genie, Best Doc), and *Act of God* (Gala Opening Night, Hot Docs). He is also director, producer, and director of photography of *Watermark* and *Black Code* (TIFF). De Pencier photographed and co-directed with Baichwal *Long Time Running,* about the Tragically Hip's historic *Man Machine Poem* tour (Gala Premiere at TIFF), which is currently in wide release. His video installation work with Baichwal has been featured at Nuit Blanche, the Art Gallery of Ontario, and the Art Gallery of Hamilton.

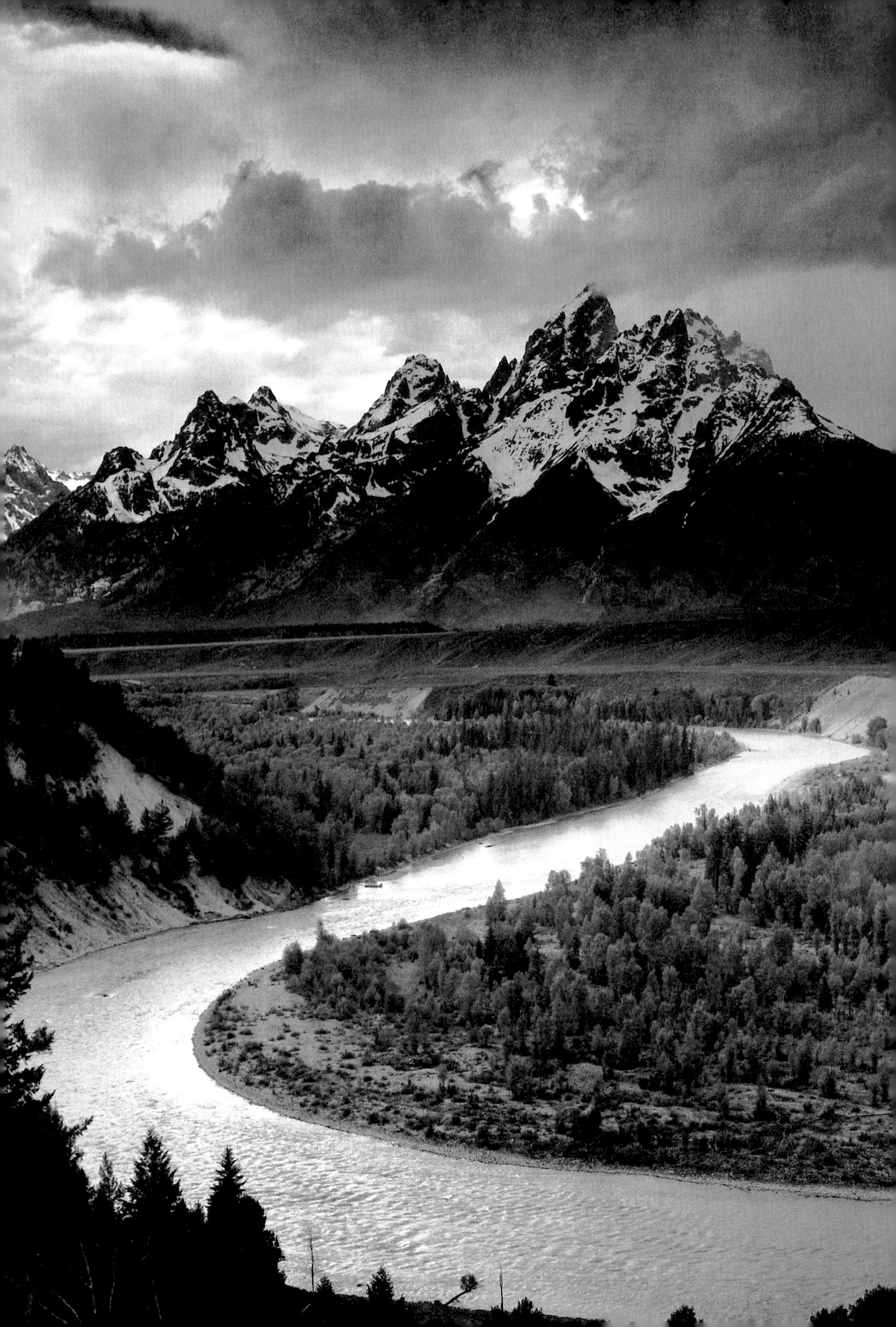

Adams, Adams, Baltz, Burtynsky: The Role of Landscape in North American Photography

Urs Stahel

In around 1970, photography, particularly in America, underwent a major transformation. The role of landscape in photography had begun to change radically. If we compare, for instance, *The Tetons and the Snake River,* taken by Ansel Adams in 1942 in the Grand Teton National Park, just south of Yellowstone National Park, with *Tract Houses, Colorado Springs* or *Pikes Peak Park, Colorado Springs,* both by Robert Adams, taken in 1968 and 1970 respectively, we instantly notice three striking differences. In the photograph by Ansel Adams, the mountain stands majestically in the centre of the image, commanding the perspectival vanishing point, while the river, in keeping with its name, seems to snake towards the viewer in the foreground. The image is dark and almost menacingly gloomy, with only occasional flashes of light appearing among the clouds above the mountain, and a glittering channel of water that seems to meander through the landscape like molten steel, breaking brightly through the shadows with a gleam that black-and-white photography so ably captures. No trace of civilization is visible in the picture. What we see spread out before us is evidently landscape at its purest—enhanced, at most, by the light, the aperture and shutter speed, and perhaps by a yellow or red filter—its romanticism dramatically heightened by subtle darkroom techniques.

The images created by Robert Adams are entirely different. They are not based on the principle of central perspective. Nor do they aim to create a centre that is filled, or even, for that matter, fulfilled. Instead, the houses sprawl unchecked into the background and seem almost to spill over the edges of the picture at both left and right. Unlike the work of Ansel Adams, these images are conspicuously, indeed almost glaringly, bright. We are not looking here at dark and dramatic atmospheric conditions in which the compositional approach and the play of light and shadow guide our mood and our emotions, but at a sunlit, brightly illuminated landscape in which the only contrast and visual structuring lies in the shadows cast by the houses themselves—the photographs having probably been taken at around four or five in the afternoon. In contrast to the emptiness and solitude portrayed by Ansel Adams, what Robert Adams depicts, first and foremost, is civilization. We find ourselves looking at a landscape and glimpsing a view of a natural world in which groups of houses or even entire settlements spread out across the plain, occupying it and taking over, as it were.

OPPOSITE PAGE
Ansel Adams, *The Tetons and the Snake River* (detail), 1942, Grand Teton National Park, Wyoming. National Archives and Records Administration, Records of the National Park Service. 79-AAG-1.

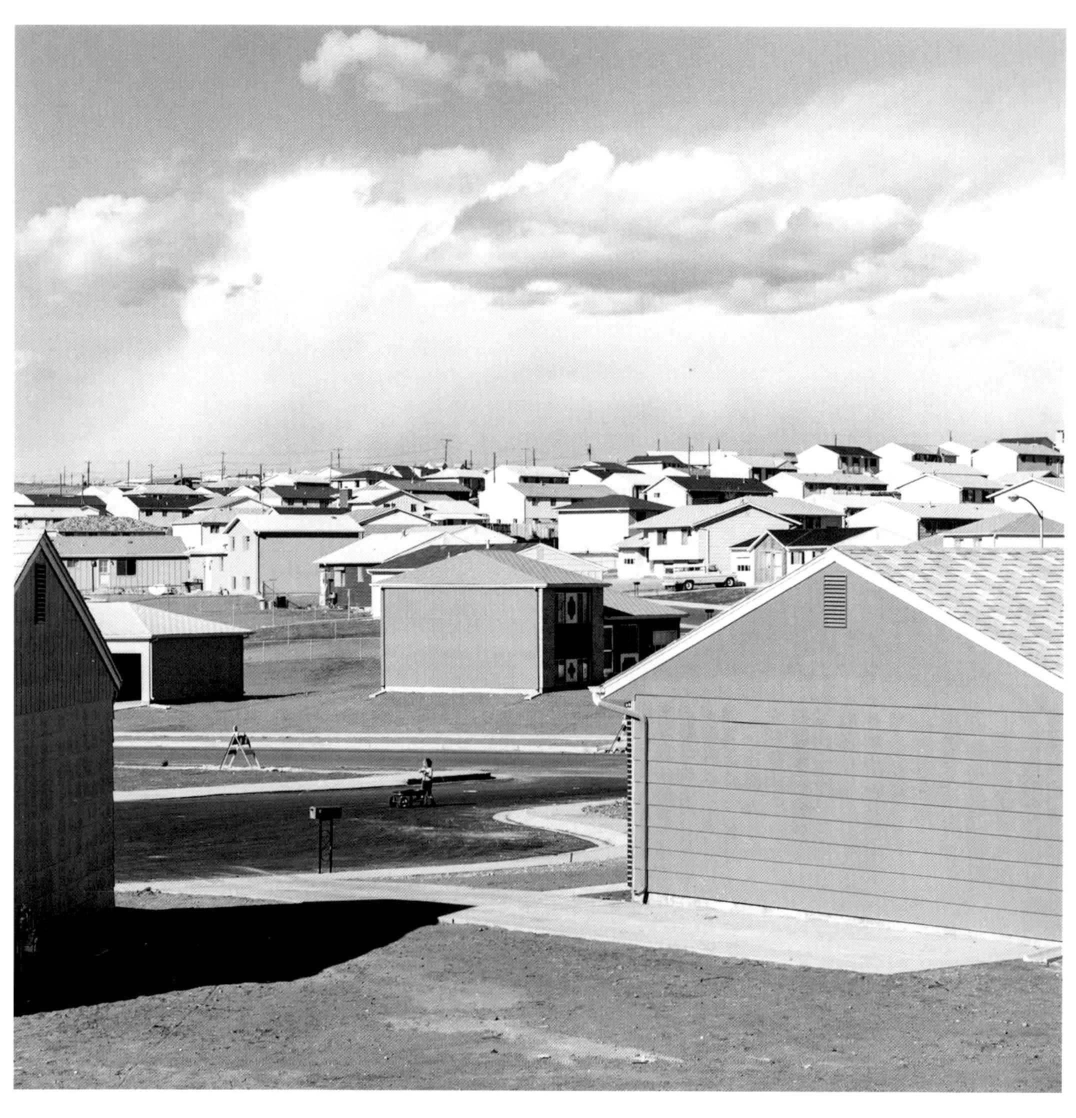

Robert Adams, *Pikes Peak Park, Colorado Springs,* 1969. Gelatin silver print, 15.24 × 15.24 cm. © Robert Adams, courtesy Fraenkel Gallery, San Francisco. 86.1972.

We can well imagine how turning 180 degrees upon our own axis could shift the gaze away from pristine, unspoiled nature and towards the ever-growing suburbs and the encroachment of the tract houses onto the plains. We can also see how this shift might trigger a sudden insight, and an awareness that the previously prevailing notion of nature in photography might not, or at least no longer, coincide with the actual reality of the American landscape. That same sudden insight, as it were, might even have been what turned an Ansel into a Robert, twenty-five years down the line. Yet this same abrupt transformation was also what inspired the work of Lewis Baltz, in addition to that of Joe Deal, Henry Wessel Jr., Frank Gohlke, Stephen Shore, and many others whom we associate with the iconic 1975 *New Topographics* exhibition that has since come to be regarded as a veritable milestone in photographic history, even though few people took note of it at the time, and fewer still found it appealing. What took place in those years was truly a double reality shift, a leap of consciousness triggered by a crack in the mirror of reality and a corresponding breach in its perception. It was akin to scales falling from the eyes, first of the photographers themselves, and then, some time later, from the eyes of those who look at photographs—in other words, a paradigm shift. On reprising the exhibition at the Center for Creative Photography in Tucson, Arizona, in 2009, curator Britt Salvesen noted how dry and unexciting some visitors had found this objective style of photography at the time.[1]

Ansel Adams was one of the last towering giants of a landscape photography that idealized and romanticized nature, and whose images depicted far more than just nature itself, by portraying beauty, the sacred, and even the godlike, as described by Estelle Jussim in *Landscape as Photograph*.[2] In his photographs, nature was transformed into a symbol. In around 1865, in a poem called "Give Me the Splendid, Silent Sun," included in his famous *Leaves of Grass,* Walt Whitman wrote:

> Give me the splendid silent sun, with all his beams full-dazzling;
> Give me juicy autumnal fruit, ripe and red from the orchard;
> Give me a field where the unmow'd grass grows;
> Give me an arbor, give me the trellis'd grape;
> Give me fresh corn and wheat—give me serene-moving animals,
>   teaching content;
> Give me nights perfectly quiet, as on high plateaus west of the
>   Mississippi, and I looking up at the stars;
> [...].[3]

Whitman was the poetic voice of a movement that was embraced, each in their own way, by photographers such as Carleton Watkins, Timothy O'Sullivan, Eadweard Muybridge, and William Henry Jackson in the nineteenth century, and by Edward Steichen, Edward Weston, Ansel Adams, and Brett Weston in the twentieth century. What their photographs have in common is that they articulate a notion of the American landscape as nature writ large: grandiose, unspoiled, beatific. Nature is idealized and stylized as the embodiment of the eternal, the enduring, the divine, challenged only by its own forces—sun, rain, snow, and wind. Landscape thus becomes a natural body that is sacrosanct and pure, in contrast to the accursed defilement that is the urban body. With that, the psyche of the American subject created, as it were, a means of escape, a refuge in the primordial, a surrogate for the divine, envisioning godliness in the natural landscape, and at the same time turning away from the gradual occupation and increasing exploitation of the land. Photography, that most fastidious instrument of truth, as it has so often been described, was deployed here not so much to reveal previously hidden truths, but rather as a form of realism that verified a poetic, pantheistic, conservationist—or even at its most extreme, political and ideological—form of whitewashing. These were symbolic exaltations sanctified by truth, because they were mechanically, technically, and chemically produced. The transcendence of any actual reality by and in photography is stated by Ansel Adams: "When words become unclear, I shall focus with photographs. When images become inadequate, I shall be content with silence."[4]

Lewis Baltz, Robert Adams, Joe Deal, Frank Gohlke, and their fellow photographers[5] took this ideal of an America that had so long been portrayed as rugged, empty, serenely beautiful, and self-fulfilled and quite simply populated it. The heroic notion of Self and Nature was transformed into Us and Our Public Park, and eventually morphed into Them and Their Picket-Fenced Gardens. Where once there was an unsullied haven of nature, there was now a row of suburban family homes, with cast-off car tires and other debris scattered around as signs indicating the presence of somebody who had already been there. Landscape thus became territory: bounded, defended, and above all, occupied. Landscape became an economic entity, to be conquered and commercialized. Lewis Baltz, in particular, turned the new image of landscape into the very antithesis of the traditionally romantic notion of nature and, in doing so, completely transformed the previously prevailing romanticism in the way images were perceived.

One notable apex in this reappropriation of nature and landscape was the series shot by Baltz in Park City, in 1979. His detailed documentation of a huge building project that involved developing a garden city and commuter hub just forty-five minutes from Salt Lake City, on the polluted grounds of a former silver mining area, marked a high point in his astute and precise observations of landscape. Park City was a large-scale construction project that the developers hoped would turn an enormous profit. It was a project of the kind that Baltz described as no longer being built for people, but being aimed solely at maximizing profits for the consortium. Over the course of two or three years, Baltz produced a portfolio of 102 photographs. It was a comprehensive and exacting visual research project, culminating in a series of small-format prints pinned to the wall that documented the construction of this gigantic hybrid in which marketing messages projecting a fictitious romantic landscape were merged with hard-nosed financial interests. Outwardly framed by all-encompassing camera views panning from the hills down toward the plain, the photographs approach the camp of the modern age step by step, encircling and capturing it according to a prepared and prescribed raster system. They first capture the material stocks, then, increasingly, the actual construction phases and, finally, the houses themselves, whereby the interiors of the gradually completed homes represent the inner promise of this complex intended as a luxury development. Baltz took on the challenge with all the precision of a surveyor, noting the location and viewpoint of each photograph—for example, *Park City, Element #40* (1979). His photographs inside the buildings are charged with ambiguity. We can never quite discern whether they are being fitted out or dismantled, erected, or demolished.

In the 1970s and 1980s, the predominant photographic attitude in art was one that sought to show things just as they were, warts and all, unromanticized, as neutral and objective as possible, viewed through a seemingly scientifically precise lens. These photographs have a deliberate air of cool detachment that eschews all human pathos and sentiment. Very much in the vein of conceptualism and structuralism, they chart structures and morphemes, exploring the typological and systematic properties of rampant urbanism, while at the same time demonstrating the geometric and economic exploitation of nature on a broad scale as an expansive functional landscape. This kind of photography could, potentially, invoke some of its predecessors of the 1920s and 1930s by referencing the works of, say, Eugène Atget, Albert Renger-Patzsch, or Walker Evans. Only the times and the signs had changed. The landscape as a whole was transformed into a vast, finite periphery—from nature to periphery, from prairie to periphery—so that photography itself has increasingly become an instrument of research.

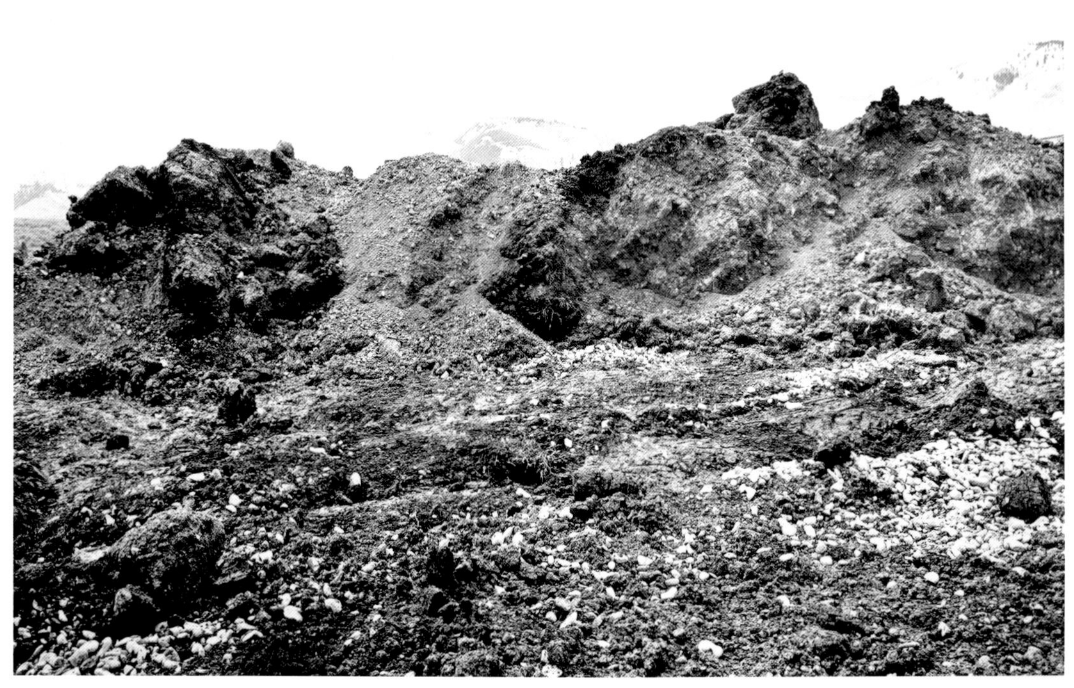

Lewis Baltz, *Park City, Element #40*, 1979. Gelatin silver print, 16.4 × 24.2 cm. The J. Paul Getty Museum, Los Angeles, Gift of Michael R. Kaplan, MD. © Estate of Lewis Baltz. 2004.168.40.

In the past two or three decades, rural and urban structures have been thoroughly shaken up once more. While the early twentieth century saw the emergence of a dynamically future-oriented vision of the city, the changes taking place today are even more fundamental. Following the conquest of space came the conquest of time—summarized by Paul Virilio as "speed"[6]—and following the conquest of time came the dissolution of place. We are here, but we work for there, and we are all connected online. What first became possible with the advent of the telephone—communicating across locations—has now become an integral part of daily life, both private and professional, through mobile networking. The relationship between the centre and the periphery has slewed 180 degrees in the opposite direction. The centres are being decanted and repurposed, while the periphery has become central to the economy. Suddenly, the periphery is everywhere, reaching all the way to the deserts, the seas, and the mountains. These are the new investment hubs and economic hotspots, cast adrift from long-developed historicity, tradition, meaning, and symbolism to become purely functional. They are zoned, sold, and used. The handling of landscape, the exploitation of nature, its economic usage and industrialization, continue apace, radicalizing at speed.

Once more, albeit from a different perspective: just as a cultural revolution swept through society in the 1960s and 1970s, today's Western industrial society is now in the throes of an ongoing high-tech revolution, in which new partnerships are forged. As Rolf Kreibich stated: "It was not the production factors of labour and capital, nor the productivity of material and energy resources, nor even the resource of information as such, that were key to social and economic structural change, but the productive factors of knowledge and technology."[7] Society and the economy are changing even faster due to the development of information and communications technology, via the internet. The revolution in information technology has recalibrated the relationships between space and time, materials and data, starkly altering spatial and temporal connections in the post-industrial era. Krishan Kumar sums this up: "Past societies [...] were primarily space-bound or time-bound. They were held together by territorially-based political and bureaucratic authorities and/or by history and tradition. Industrialism confirmed space in the nation state while replacing the rhythms and tempo of nature with the pacing of machine. [...] The computer, the symbol of the information age, thinks in nanoseconds, in thousandths of microseconds. Its conjunction with the new communications technology thus brings in a radically new space-time framework for modern society."[8]

It is into the reality of this new, radicalized world that Edward Burtynsky inserts his photography. His work addresses how we are changing the world we live in and the environment we inhabit, not just locally but globally. We are impacting our natural resources with unprecedented intensity and speed in an era already labelled the Anthropocene. We are doing so according to the concept of an epoch shaped by humans, beginning with the 1800–1945 phase of industrialization, and followed by the great acceleration of the postwar era, then further compounded by a global population explosion in combination with widespread sealing of our landscapes through megacities, economic globalization, and the rise of an excessively consumerist society. There is now a growing awareness of the consequences of our impact on the global ecosystem.[9] Sahel Syndrome, Overexploitation Syndrome, Dustbowl Syndrome, Burnt Earth Syndrome, Aralsee Syndrome, Favela Syndrome, Disaster Syndrome, Smokestack Syndrome, Landfill Syndrome: these are just a few of the terms used to describe some of the typically undesirable side effects and environmentally damaging patterns of the natural and civilizational trends that form part of the great shadow being cast by humankind over the face of the earth's ecosystem.[10]

Like all contemporary photographers, Burtynsky is up against the increasing invisibility and disappearance of essential elements, the interplay of cause and effect: cables, black boxes, data banks, abstractions. His strategy must necessarily diverge from that of the *New Topographics*. Both the idea and the reality of the man-altered landscape have now progressed so far that it is no longer possible to approach these problems by way of visual fieldwork alone. Instead, a symphonic visual world needs to be composed and presented in a way that will seduce and shock the viewers and jolt them into awareness, leaving them with guilty consciences. Burtynsky does this in several ways. For instance, he may elevate an image into the realms of heraldic syntax by documenting a theme visually while at the same time heightening it into a symbol. Often, Burtynsky will clamber into a helicopter, or fly a drone, in order to capture and show broad swathes of landscape from above, like some animated map that can render the increasing appropriation of nature. The landscapes in his photographs often look like rolled-out sheets of reptile skins onto which human usage has been stamped or etched. We humans seem to plow over the skin of the earth, tattooing it and gradually, bit by bit, at times even explosively, throwing it off-kilter. Burtynsky counters this entwined network of increasingly denser and constantly proliferating interventions into nature by means of the sheer power of his images and his focus on symmetries, circles, grids, and sharply drawn geometric lines. He unleashes a crescendo of colours and forms in large-scale images, which, together with the filmic input of Nicholas de Pencier and Jennifer Baichwal—in their carefully chosen angles and cuts, combined with searing, compelling slow motion—present us with a visual array that is as potently impossible for the viewer to resist as it would be for the listener to withstand Beethoven's Fifth Symphony. He conjures visual worlds that meld inside with outside, here with there, the up-close with the beyond, churning up our thoughts and emotions. The effect is further heightened by starkly contrasting the wonders, origins, and indescribable beauty of nature with the augmented reality (AR) technique of reconstituting a lost world, as in the images of burning elephant tusks that send a message of protest against poachers.

All these events, influences, impacts, interventions, transformations, and imprints are addressed in scientific and political discourse under the term "Anthropocene," by which we acknowledge that, following the Holocene era, we have now entered an epoch—albeit very recently in terms of overall time—in which humankind is changing the world and shaping it in ways that are accelerating the extinction of species, the pollution of the food chain with microplastics, and the radioactive contamination of substances and the air that we breathe, such that we are truly "consuming the planet to excess"[11] to an extent that was previously achieved only by geological phenomena occurring over the course of thousands, hundreds of thousands, or even millions of years.

There is, however, one important caveat: it is not some value-neutral humankind that is pushing this exploitation and destruction. It is important to state that "neither the individual nor humankind in general has become a geohistorical power, but very specific individuals who have inserted themselves into the social and economic structures of the OECD [Organisation for Economic Co-operation and Development] world and who have burdened the entire world populace with a kind of guilt by association for problems such as climate change, which are, in truth, caused by a minority in the capitalist western world."[12] The Anthropocene era, according to critics of the term, is a result of the actions of powerful protagonists in the global economy and politics of both "old" and "new" imperialism.[13]

While the term itself may be debatable,[14] the impact of human intervention is not. Photographic, artistic, literary, and cultural works have become just as central to raising awareness and fostering discussion regarding this difficult situation as any statistics and reports. Edward Burtynsky, Jennifer Baichwal, and Nicholas de Pencier do incredibly important work in this field. Landscape may have lost its innocence and the sacredness ascribed to it, but on the other hand, it is our own actions and exploitation that have overstepped all previously imaginable bounds.

Urs Stahel is the curator of Fondazione MAST and the advisor to its collection of industrial photography, and the co-curator of the *Anthropocene* exhibition.

1. Anthony Bannon, director of the George Eastman Museum until 2012 and previously a filmmaker and journalist for *Buffalo News,* recalls having found the show uninspiring, as reported by Alison Nordström in her contribution to the history of the reception of *New Topographics,* in the exhibition catalogue issued by the Center for Creative Photography at the University of Arizona under the auspices of then director Britt Salvesen. Landscape photographer Mark Klett, who was enrolled at the Visual Studies Workshop in Rochester at the time, said that the photos seemed "intentionally boring," and that he did not like how the photographers detached themselves from their motifs. He considered that the show's underlying concept of objectivity had not been achieved. See Britt Salvesen, *New Topographics* (Göttingen: Steidl, 2009).

2. See Estelle Jussim and Elizabeth Lindquist-Cock, *Landscape as Photograph* (New Haven: Yale University Press, 1985).

3. Walt Whitman, "Give Me the Splendid, Silent Sun," *Walt Whitman Archive*, https://whitmanarchive.org/published/LG/1867/poems/184.

4. Ansel Adams, *AB Bookman's Weekly* 76, nos. 19–27 (1985): 3,326.

5. Britt Salvesen, *New Topographics* (Göttingen: Steidl Verlag, 2009).

6. Paul Virilio, *Speed and Politics,* trans. Mark Polizotti (Los Angeles: Semiotext(e), 2006). Originally published as *Vitesse et Politique* (Paris: Edition Galilee, 1977).

7. Rolf Kreibich, *Die Wissensgesellschaft* (Frankfurt: Suhrkamp Verlag, 1986).

8. Krishan Kumar, *From Post-industrial to Post-modern Society: New Theories of the Contemporary World* (Malden: Blackwell, 2005).

9. Hans Gebhardt, "Das Anthropozän—zur Konjunktur eines Begriffs," article 3 in *HDJBO*, 1 (2016): 28.

10. Ibid., 33–34.

11. John Urry, "Consuming the Planet to Excess," *Theory, Culture & Society* 27, nos. 2–3 (2010): 191–212.

12. C. Schwägerl and R. Leinfelder, "Die menschengemachte Erde," *Zeitschrift für Medien-und Kulturforschung* 5, no. 2 (2014): 233–240.

13. David Harvey, *The New Imperialism* (Oxford: Oxford University Press, 2003).

14. For a discussion on this term, see Jürgen Manemann, *Kritik des Anthropozäns—Plädoyer für eine neue Humanökologie* (Bielefeld: Transcript Verlag, 2014). The German theologian, philosopher, and director of the Hanover Institute of Philosophical Research (Forschungsinstitut für Philosophie Hannover, FIPH) —a critic of the Anthropocene—outlines the relationship between humans and the environment. According to him, "Humans shape nature. That is the core of the Anthropocene thesis." He criticizes extropianism and transhumanism, both of which embrace the tenet of transcending the human. Furthermore, he argues that our current challenges can be overcome in more pragmatic and even political ways, and that we should take a less hubristic approach to the world.

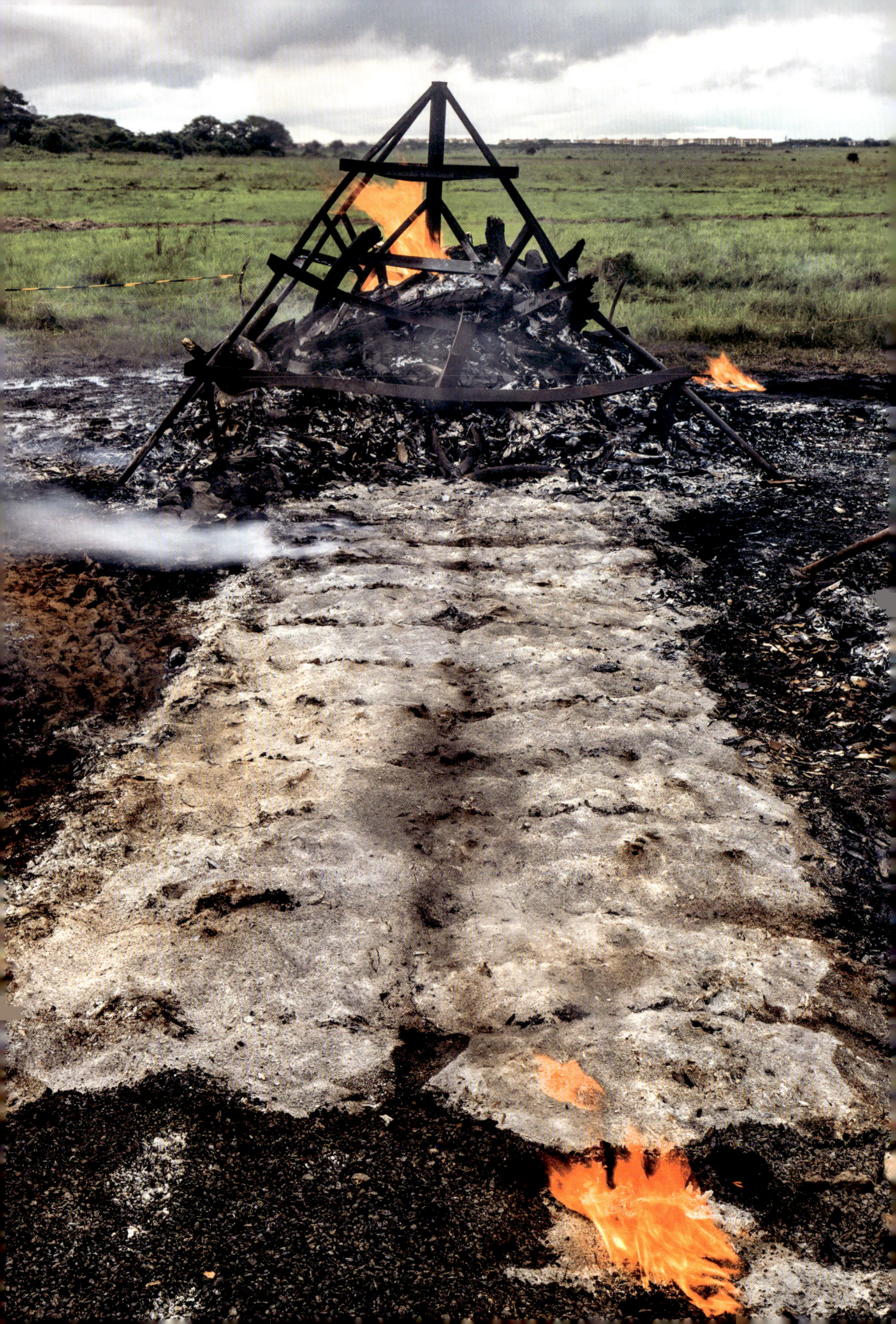

The Art Museum and the Anthropocene

Andrea Kunard

The Anthropocene as a Curatorial Approach

I understand the exhibition *Anthropocene* as an indecisive moment in the art museum. Because it is titled by a term impossible to define,[1] the exhibition does not contain or restrict meaning. Rather, it assumes a role of maintaining openness and dialogue. In this sense, "Anthropocene" operates as a contingent or provisional universal—an expression of something occurring widely, but experienced differently, depending on locale or individual circumstances. Its plural nature provides a forum for multidisciplinary approaches.[2] As a process, it assembles, breaks apart, and reconfigures boundaries, mixing visual art with science, literature, activism, and theory. Most importantly, as a protean concept, Anthropocene positions the art museum as a space and forum for discussion, providing moments of community for the exchange and debate of ideas, while capitalizing on the power of aesthetics to hold the viewer in a place of reflection.

Acting as a placeholder term for discussion, "Anthropocene" remains open to its etymology, recognizing the consequences of human actions on the planet, without equating all humans equal in their destructiveness.[3] It challenges the buttresses of human exceptionalism, or the idea that the human species is distinct from others, and dives deeper to challenge the idea of "Human"[4] and "Species" as universal terms. In its focus on an environmental crisis of planetary proportions, one that will consume present and future states of existence (of all living species), Anthropocene combines natural and human histories; it refashions absence of history and continuity as a geological epoch, where *Ánthrōpos* effects are "written in stone." Human activity, traced as isotopes, microplastics, concrete, carbon dioxide, and black carbon, merge with the natural force of the geological that shapes continents and tips land masses. The original date of this planetary crisis is still in dispute, and perhaps can be dispensed with in order to highlight the lack of need for origins. Humans and the residues of their activities have been enfolded within the planetary crust for millennia. For proponents of the Anthropocene, however, events in recent times such as nuclear explosions, decreased biodiversity, massive extinctions, and excesses associated with petroculture, including increased carbon emissions and plastic production, comprise the limits of the Holocene and the move into our Anthropocenic present.[5]

OPPOSITE PAGE
Edward Burtynsky, *Burned Ivory Tusks #2, May 1, Nairobi, Kenya* (detail), 2016.

With the Anthropocene, we recognize the futility in pitting disciplines against one another in a binary fashion. Science and art mingle; human, vegetable, and animal become intimately intertwined. As French philosopher and sociologist Bruno Latour says, "Because of the very logic of the Anthropocene, you are inserted into the phenomena you study in a way that is unexpected and still unfathomed. The idea of a science that emerges from the dispassionate study of external phenomena is now much more difficult to sustain. The very distinction between the social and natural sciences breaks down, because the argument that you are not supposed to be involved in what you study can no longer be maintained."[6]

For art, the Anthropocene represents an opportunity to become immersed in the sensate and conceptual, and to trace their unexpected merging and configurations. One goal is to draw attention to the power of lens-based media, and how they, in their continuous reconfiguring of reality, provide new sensorial experiences, refashion memory and history, and reposition viewers with novel and sometimes multiple viewpoints. Unexpected understandings of reality may occur; photography is particularly malleable in the Anthropocene. Digital renderings of the world in hyper-saturated colour and high contrast may appear hyperreal to present-day viewers, and totally normal to future ones. Algorithmically driven generative adversarial network (GAN) imagery—images created solely by computers—may be deemed as "real" as those captured through photographic lenses, and operate as substitutes to what actually exists in the world.[7] To explore the role of photography and other lens-based media in the Anthropocene means being open to their contradictory status[8] where, especially through the plasticity of the digital, photographic practices and products enter larger contexts that merge fantasy and reality, fiction and truth. As well, it must be acknowledged that the very tools of inquiry into the Anthropocene—cameras, lenses, computers, scanners, and printers—eventually contribute to the problem under scrutiny when they enter technofossil middens along with other products of consumer capitalism.

Lens-based media in the Anthropocene decentre the subject by amalgamating expectations, points of view, information, sensation, and objective and subjective positions. The camera's traditional function is to record and heighten the human capacity to scrutinize the world. As a prosthesis to vision, the camera merges biology with technology. This quality is amplified in the Anthropocene. The camera is now not a stand-alone device: it combines with other technologies, such as satellites, rockets, drones, planes, and helicopters, to offer a transcendent vision of reality. When wedded to the internet, camera renderings of individual, momentary experiences are made available to innumerable others at lightspeed rates of exchange, intermingling personal, private, communal, and public states. Images are data to be mined, quantified, and reconfigured into new matrices of knowledge. Information proliferates exponentially, its value linked to the volume of traffic it attracts.

The Anthropocene and the Art Experience

For the viewer in the art gallery, multiple sensory and temporal experiences are made available through the dynamics of photographic and filmic aesthetics. In a basic sense, time is both an essential element of the still image, which is generated in a fraction of a second, and the material state of what is depicted. As well, the click of the shutter freezes subject matter in time, exorcizing it from the flow and flux of reality, a condition that may raise concern as to its continuity. In the exhibition, these anxieties are related to the environment and subject matter under threat, as seen in Edward Burtynsky's murals *Cathedral Grove #1, Vancouver Island, British Columbia, Canada* and *Pengah Wall #1, Komodo National Park, Indonesia*, the former depicting an old-growth forest and repository of innumerable fragile ecosystems, the latter an ecosystem of coral reefs fashioned through the incremental, delicate creation of marine invertebrate communities. A more humbling vision of time, frozen as natural formation, is also seen in the depiction of stratified rock caching a potential geological "Golden Spike" moment in *Basque Coast #1* and *#2*. The photograph's privileged relation to evidence also informs experiences of time. There are images depicting the impact of long-term terraforming projects, such as farming practices in Monegrillo, Spain, photographically fashioned into crazy dance squiggles of roads, pathways, and terraces. Oppositely, the voraciousness of a gargantuan machine in Hambach, Germany, scooping up landscapes indicates efficiency-driven activities—ones that, because governed by the adage "Time is money," incur both horror and admiration for their intertwining of human destructiveness and inventiveness.

Aesthetics also play a significant role in heightening sensorial and temporal experiences. In many of Burtynsky's works, a lack of horizon accentuates the image's two-dimensional formal qualities, transfiguring specific, geographical elements into visual puzzles. The predominant temporal experience is one of pause created by the force of the image that demands concentrated viewing. Experience wavers between the sensuality of pure aesthetics, manifested in the play of patterning, texture, and colour, and the intellectual, engaged through photography's descriptive, documentary qualities, where sinuous lines reformulate as roads, cubes reconfigure as buildings, and amorphous shapes transform into tailing ponds.

Throughout the *Anthropocene* exhibition, the still image is countered by the movement of filmic components. Although Jennifer Baichwal and Nicholas de Pencier often work closely with Burtynsky on the same projects, the relation of film to photography is not necessarily congruous. The two mediums maintain a fundamental dichotomy that once prompted curator and writer David Campany to ask, "What *is* the movement of film and what *is* the stillness of photography? Is it that the film image changes over time while the photograph is fixed? Not exactly. That photographs are *about* stillness and films *about* movement? Possibly, but that still misses something."[9] He observes that throughout the early part of the twentieth century, photographers and filmmakers did not reflect on the distinction until Andy Warhol distinguished cinematic duration from depicted movement in *Sleep* (1963). Within *Anthropocene*, cinematic duration vies with documentary narrative, providing viewers with different experiences of the subject under scrutiny. The film *Gotthard Base Tunnel, Gotthard, Switzerland*, is decidedly durational, and one way to draw viewers into the gallery. It also serves to disorient, emphasizing the importance of physical sensations when encountering a work of art. As well, the film celebrates speed and the technology that provides the capacity for humans, and the gallery viewer by extension, to move quickly through mountains. The work also indicates how the cinematic lens constructs a filmic vocabulary through focus on the formal quality of subject matter. The very essence of the tunnel is a means to explore duration. The filmic quality of time length also serves to indicate the sheer volume of resource extraction, as in the scene that depicts coal being transported with Fordian efficacy in *Coal Trains, Wyoming, USA*—the vanishing point effectively used to express the ceaseless appetite of late capitalist economies. Another cinematic technique, the panning shot, sensuously complements the presentation of riband geological subject matter in *Hambach Lignite Mine, Germany*. In *Coral Bleaching, Great Barrier Reef, Australia*, DSLR time lapse not only condenses viewing time, but, along with the photographic technique of focus stacking, allows viewers to appreciate animate appetite, and the sticky, fleshy, ingurgitating, esurient, entrapping protuberances of life that flourish at the edges and maintain the possibilities of always becoming. This intimate and intense vision, however, is undercut by the lamentable process of bleaching caused by pollution and temperature swings in oceans that leaves coral vulnerable and seemingly ossified.

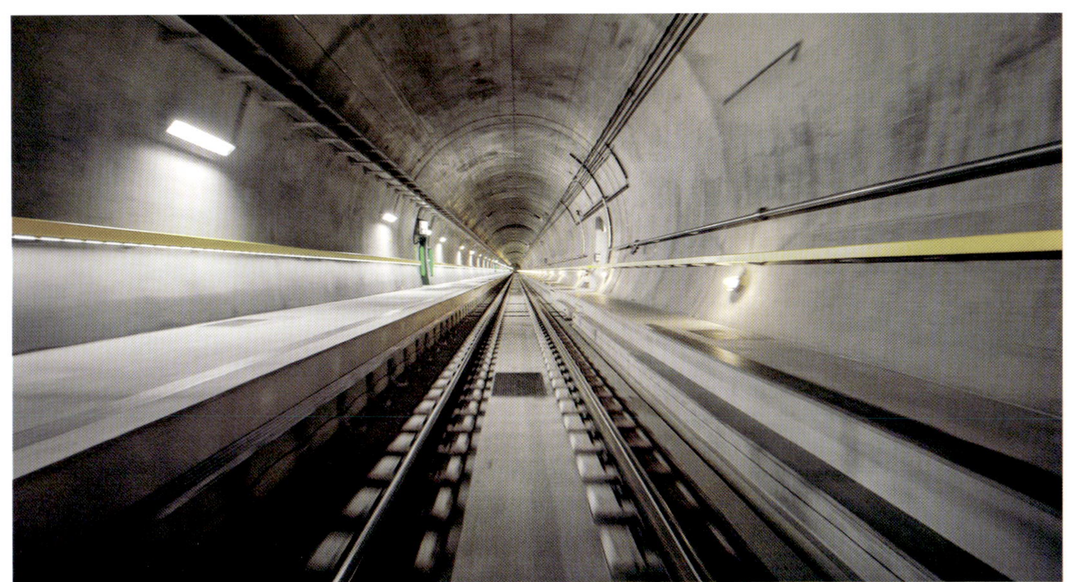

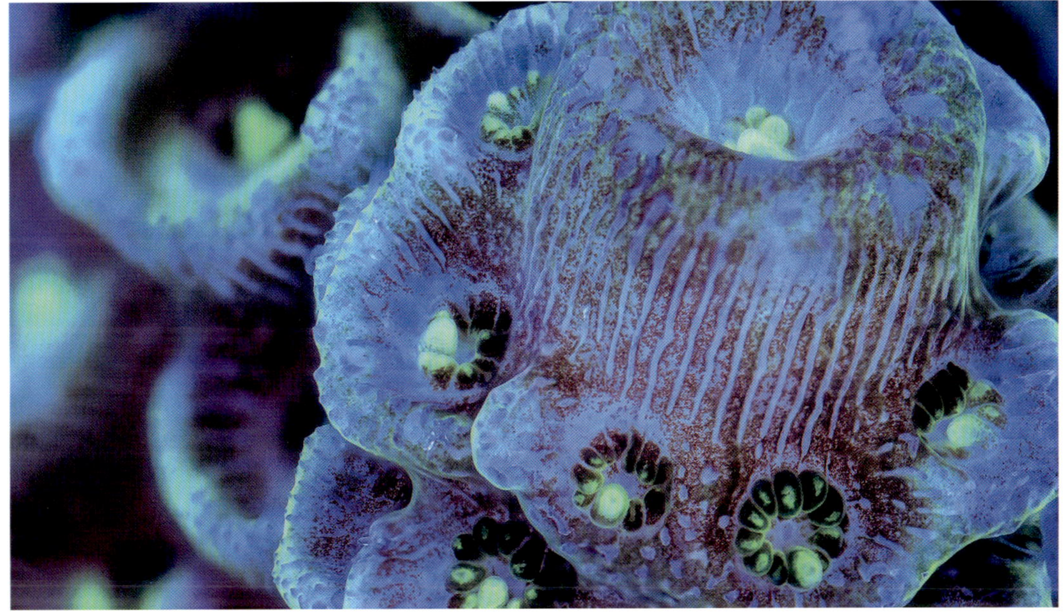

TOP
Jennifer Baichwal and Nicholas de Pencier, *Gotthard Base Tunnel, Gotthard, Switzerland* (film still), 2018.

BOTTOM
Jennifer Baichwal and Nicholas de Pencier, *Coral Bleaching, Great Barrier Reef, Australia* (film still), 2018.

Formal qualities also inform experience through the visual echoes at play throughout the exhibition that merge human and natural formations. Embankments in the open-pit mine of *Morenci Mine #1, Clifton, Arizona, USA* configure as giant leaves, their larger surroundings visually suggestive of marble patterning. In *Chino Mine #5, Silver City, New Mexico, USA*, the mine takes on the aspect of a wound with a black centre, encircled by rubious markings suggestive of inflammation. Or, its ringed patterns of terraces and roadways recall the cross section of a tree exhibiting riotous fluctuations of annual growth. Natural and human-made environs also visually intermingle. The vast ecosystem of colourful coral polyps and marine life resonate with the vibrant synthetic environment created by mounds of plastic bottles in the landfill at Dandora, Kenya.

As much as photographic and filmic works in *Anthropocene* are strongly configured through visual means, the documentary mode is also powerfully at play, offering viewers the possibility of imaginatively engaging with imagery as well as with subject matter. The very scale of the images matches the immensity of the issues under investigation. Burtynsky presents disturbing subject matter in a visually alluring manner to capture the viewer's attention and encourage thoughtfulness.[10] Industrial sites and machinery inspire an appreciation for human achievement, and concern that consumption of resources is unsustainable.[11] Baichwal and de Pencier's filmic presentations offer narrative possibilities and empathetic engagement with subjects in their particular context; viewers can accompany a Dandora landfill resident into synthetic canyons of plastic materials and polymer strata, reflecting on the relation of their own lifestyles to such landscapes. The inferno depicted in *Elephant Tusk Burn, Nairobi National Park, Kenya* positions viewers as witnesses to an action of extreme prejudice against a particular species, but it is also a reminder that the event results from notions of human exceptionalism that declare Human as distinct from nature and other species, and thus entitled to maintain dominion over all. As writer Donna Haraway notes, "If we appreciate the foolishness of human exceptionalism, then we know that becoming is always becoming *with,* in a contact zone where the outcome, where who is in the world, is at stake."[12]

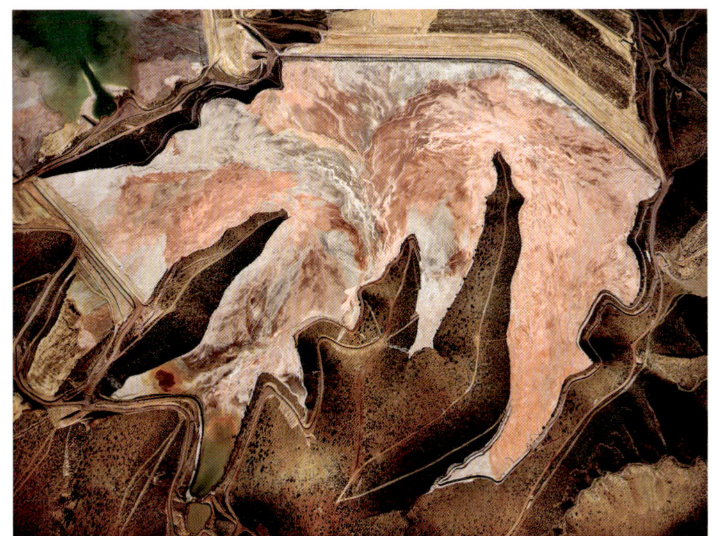

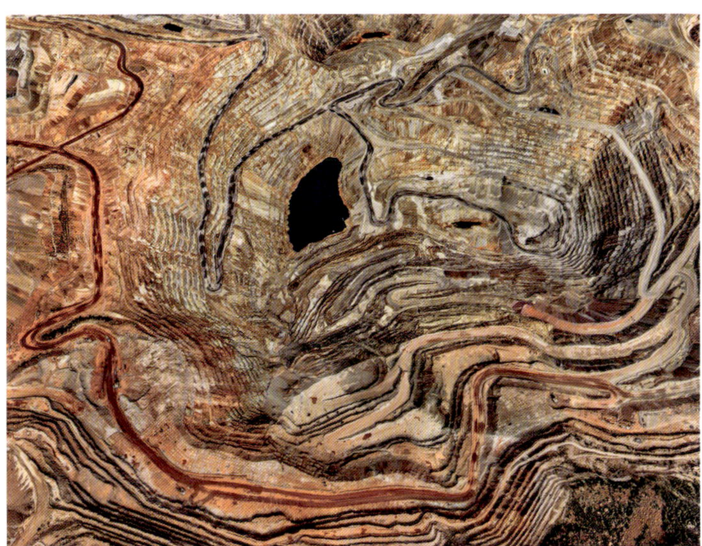

TOP
Edward Burtynsky, *Morenci Mine #1, Clifton, Arizona, USA*, 2012.

BOTTOM
Edward Burtynsky, *Chino Mine #5, Silver City, New Mexico, USA*, 2012.

The possibilities of "becoming with" are granted, in *Anthropocene,* through augmented reality installations that merge familiar dichotomies such as human/technology, subject/object, animal/human, presence/absence, art/science, and nature/culture. The experience provides a means to break open, reorganize, and reconfigure seemingly oppositional stances to ones that encourage openness, intermingling, and curiosity. The technological engagement with elephant tusks, isolated Douglas fir trees, and a near extinct species of rhino may invoke a range of responses—emotional, intellectual, inquisitive, confused. It may also provide possibilities of maintaining openness to diffractive patterns of encounter, not merely those that are singular reflections of one's values.[13] The virtual space of the installation is a public arena of exploration and interrogation, of thoughtfulness and encounter with innumerable Others, of thinking about where and why we draw certain boundaries, of what is included and excluded, and of the ethical consequences of such decisions.

Andrea Kunard is Associate Curator of the Canadian Photography Institute, National Gallery of Canada and co-curator of the *Anthropocene* exhibition.

1. The topic of the Anthropocene has generated volumes of literature, with many, sometimes, contesting theories. For more information, visit the Anthropocene Working Group website, https://theanthropocene.org/topics/anthropocene-working-group/.

2. Dipesh Chakrabarty states that we cannot assume a detached position with respect to such pressing issues as global warming, because all of humanity is affected in different and unpredictable ways. He proposes a universal based on a "shared sense of catastrophe" that "calls for a global approach to politics without the myth of a global identity." He terms this a "negative universal history." Dipesh Chakrabarty, "The Climate of History: Four Theses," *Critical Inquiry* 35, no. 2 (Winter 2009): 221–222. See also commentary by Bruno Latour, note 6.

3. For an Indigenous perspective on the Anthropocene, see Heather Davis and Zoe Todd, "On the Importance of a Date, or Decolonizing the Anthropocene," *ACME: An International Journal for Critical Geographies* 16, no. 4 (2017): 761–780.

4. Researchers have discovered that only 43 percent of our body cell count is human. The rest belongs to microscopic creatures such as bacteria, viruses, fungi, and archaea. James Gallagher, "More Than Half Your Body Is Not Human," *BBC News*, April 10, 2018, http://www.bbc.com/news/health-43674270.

5. Numerous names have been proposed for this present moment—Anthropocene, Plantationocene, and Capitalocene, to name a few. Kim Stanley Robinson further understands our time as that of "The Dithering" or a "state of indecisive agitation." Donna Haraway, "Anthropocene, Capitalocene, Plantationocene, Chthulucene: Making Kin," in *Environmental Humanities* 6 (2015): 161.

6. Bruno Latour, "Diplomacy in the Face of Gaia," in *Art in the Anthropocene*, ed. Heather Davis and Etienne Turpin (London: Open Humanities Press, 2015), 44.

7. See "Microsoft Researchers Build a Bot That Draws What You Tell It To," January 18, 2018, https://blogs.microsoft.com/ai/drawing-ai/. I am indebted to the research of Dr. Estelle Blaschke and her paper "When Images Become Data," delivered at the symposium *Photography: The Black Box of History*, held at Ryerson University, Toronto, Canada, March 16, 2018.

8. Heather Davis and Etienne Turpin, "Art & Death: Lives Between the Fifth Assessment & the Sixth Extinction," *Art in the Anthropocene*, 14.

9. David Campany, *Photography and Cinema* (London: Reaktion Books, 2008), 24. Emphasis in original.

10. As Burtynsky has stated about his method, "I have you—you're reacting to the work, you're engaging and [...] it's far more interesting to me to work with that tension than to say, 'Here's something that supports the idea that this is bad.'" "Edward Burtynsky on *Oil*," *CBC News*, Arts & Entertainment, last modified April 6, 2011, http://www.cbc.ca/news/arts/story/2011/04/05/edward-burtynsky.html.

11. Ibid.

12. Donna Haraway, *When Species Meet* (Minneapolis: University of Minnesota Press, 2008), 244.

13. See Karen Barad on ideas of refraction and entanglement that provide nonconfrontational modes of engagement: "Reading insights through one another in attending to and responding to the details and specificities of relations of difference and how they matter." Karen Barad, *Meeting the Universe Halfway: Quantum Physics and the Entanglement of Matter and Meaning* (Durham: Duke University Press, 2007), 71.

Notes (pages 60–187)

The spelling of "tons" (or "tonnes") in the texts reflects usage in the original sources, where "tonne" generally refers to the metric ton.

Page 60

1. "Flysch Formation in Zumaia," *Amusing Planet*, June 8, 2015, http://www.amusingplanet.com/2015/06/flysch-formation-in-zumaia.html.
2. Joan Poch and Jon Paul Llordés, "The Basque Coast Geopark: Support for Good Practices in Geotourism," *GeoJournal of Tourism and Geosites* 8, no. 2 (2011): 273.
3. Allison A. Baczynski, Francesca A. McInerney, Scott L. Wing, Mary J. Kraus, Jonathan I. Bloch, and Ross Secord, "Constraining Paleohydrologic Change During the Paleocene-Eocene Thermal Maximum in the Continental Interior of North America," *Palaeogeography, Palaeoclimatology, Palaeoecology* 465 (2017): 237–246.
4. Laia Alegret, Silvia Ortiz, Xabier Orue-Etxebarria, Gilen Bernaola, Juan I. Baceta, Simonetta Monechi, Estibaliz Apellaniz, and Victoriano Pujalte, "The Paleocene–Eocene Thermal Maximum: New Data on Microfossil Turnover at the Zumaia Section, Spain," *Palaios* 24, no. 5 (2009): 318–328.
5. Birger Schmitz, Victoriano Pujalte, M. Eustoquio, Simonetta Monechi, Xabier Orue-Etxebarria, Robert P. Speijer, Laia Alegret, et al., "The Global Stratotype Sections and Points for the Bases of the Selandian (Middle Paleocene) and Thanetian (Upper Paleocene) stages at Zumaia, Spain," *Episodes* 34, no. 4 (2011): 220–243.
6. "The Development of Agriculture," *Genographic Project*, accessed April 5, 2018. https://genographic.nationalgeographic.com/development-of-agriculture/.

Page 62

1. "Kenya: Waste Dump Poses Health Hazard to Children, UN Agency Warns," United Nations, accessed April 3, 2018, https://news.un.org/en/story/2007/10/235002-kenya- waste-dump-poses-health-hazard-children-un-agency-warns.
2. "Buried in Dandora: Voices of Nairobi's Waste Management Disaster," Pulitzer Center, January 5, 2017, https://pulitzercenter.org/projects/kenya-nairobi-dandora-waste-management-public-health-poverty-sanitation-crisis.
3. Anne O'Mahoney, "Trash and Tragedy: The Impact of Garbage on Human Rights in Nairobi City," Concern Worldwide, https://www.concern.net/sites/default/files/resource/2012/09/5808-trash_and_tragedy-final.pdf.
4. Jan Zalasiewicz, Colin N. Waters, Juliana A. Ivar do Sul, Patricia L. Corcoran, Anthony D. Barnosky, Alejandro Cearreta, Matt Edgeworth, et al., "The Geological Cycle of Plastics and Their Use as a Stratigraphic Indicator of the Anthropocene," *Anthropocene* 13 (2016): 5.
5. Ibid.
6. Ibid.

Page 66

1. David Conrad, "Dandora: Conflicting Views and Multiple Struggles," Pulitzer Center, March 28, 2012, https://pulitzercenter.org/reporting/dandora-conflicting-views-and-multiple-struggles.
2. Micah Albert, "Kenya Poor Cling to Dump Site," Pulitzer Center, April 30, 2012, https://pulitzercenter.org/reporting/kenya-poor-cling-dump-site.
3. David Conrad, "Nairobi's Garbage Dump Pits Pickers Against Neighbours," *Globe and Mail,* May 13, 2012, https://www.theglobeandmail.com/news/world/nairobis-garbage-dump-pits-pickers-against-neighbours/article4170400/.
4. "Kenya: Waste Dump Poses Health Hazard to Children, UN Agency Warns," United Nations, October 5, 2007, https://news.un.org/en/story/2007/10/235002-kenya-waste-dump-poses-health-hazard-children-un-agency-warns.
5. Ibid.

Page 68

1. C. N. Waters, J. Zalasiewicz, C. Summerhayes, A. D. Barnosky, C. Poirier, A. Gałuszka, A. Cearreta, M. Edgeworth, E. C. Ellis, M. Ellis, C. Jeandel, R. Leinfelder, J. R. McNeill, D. deB Richter, W. Steffen, J. Syvitski, D. Vidas, M. Wagreich, M. Williams, A. Zhisheng, J. Grinevald, E. Odada, N. Oreskes, and A. P. Wolfe, "The Anthropocene Is Functionally and Stratigraphically Distinct from the Holocene," *Science* 351, no. 6269 (2016).
2. Zhijun Ma, David S. Melville, Jianguo Liu, Ying Chen, Hongyan Yang, Wenwei Ren, Zhengwang Zhang, Theunis Piersma, and Bo Li, "Rethinking China's New Great Wall," *Science* 346, no. 6212 (2014): 912.
3. J. P. M. Syvitski and A. Kettner, "Sediment Flux and the Anthropocene," *Philosophical Transactions of the Royal Society A: Mathematical, Physical and Engineering Sciences* 369, no. 1938 (2011): 957–75.
4. Ibid.
5. Ibid.
6. Syvitski and Kettner, 2011.
7. Zhijun et al., 2014.
8. Ibid., 912.

Page 72

1. John Vidal, "Here Come the Megacities," *Ensia*, January 19, 2018, http://edge.ensia.com/here-come-the-megacities.
2. Ibid.
3. Tolu Ogunlesi and Andrew Esiebo, "Inside Makoko: Danger and Ingenuity in the World's Biggest Floating Slum," *Guardian*, February 23, 2016, https://www.theguardian.com/cities/2016/feb/23/makoko-lagos-danger-ingenuity-floating-slum.
4. Mimi Onuoha, "A 5-Mile Island Built to Save Lagos's Economy Has a Worrying Design Flaw," *Quartz Africa*, March 18, 2017, https://qz.com/923142/the-flaw-in-the-construction-of-eko-atlantic-island-in-lagos.
5. Martin Lukacs, "New, Privatized African City Heralds Climate Apartheid," *Guardian*, January 21, 2014.
6. Alexis Okeowo, "A Safer Waterfront in Lagos, If You Can Afford It," *The New Yorker,* August 20, 2013.
7. "World Population Prospects: The 2017 Revision | Multimedia Library—United Nations Department of Economic and Social Affairs," United Nations, June 21, 2017, https://www.un.org/development/desa/publications/world-population-prospects-the-2017-revision.html.
8. Ibid.

Page 80

1. "The Great Sprawl of China," *The Economist* (U.S.), January 24, 2015.

Page 82

1. Tom Uhler, "Want an Idea of How Many Cars Were Damaged in Harvey? Get a Birds-eye View," *Star-Telegram*, October 16, 2017, http://www.star-telegram.com/news/state/texas/article179175856.html.
2. "Fast Facts: Hurricane Costs," Office for Coastal Management, National Oceanic and Atmospheric Administration, accessed April 28, 2018, https://coast.noaa.gov/states/fast-facts/hurricane-costs.html.
3. "U.S. Energy Information Administration—EIA—Independent Statistics and Analysis," Gulf of Mexico Fact Sheet—Energy Information Administration, accessed April 4, 2018, https://www.eia.gov/special/gulf_of_mexico/.
4. Ibid.
5. Ernest Scheyder and Erwin Seba, "Harvey Throws a Wrench into U.S. Energy Engine," *Reuters*, August 28, 2017, https://www.reuters.com/article/us-storm-harvey-energy/harvey-throws-a-wrench-into-u-s-energy-engine-idUSKCN1B70YQ.
6. Steve Mufson, "ExxonMobil Refineries Are Damaged in Hurricane Harvey, Releasing Hazardous Pollutants," *Washington Post*, August 29, 2017.
7. Michael Isaac Stein, "After Harvey, What Will Happen to Houston's Oil Industry?" *Wired*, September 21, 2017.

Page 84

1. M. Yúfera and A. M. Arias, "Traditional Polyculture in 'Esteros' in the Bay of Cádiz (Spain)," *Aquaculture* 35 (2010): 3.
2. Ibid., 23.
3. Ibid.
4. Ibid.

Page 88

1. John Fleck, *Water Is for Fighting Over: And Other Myths about Water in the West* (Washington, D.C.: Island Press, 2016).
2. Felicity Barringer, "Empty Fields Fill Urban Basins and Farmers' Pockets," *New York Times*, October 24, 2011.
3. Glenn D. Schaible and Marcel P. Aillery, *Water Conservation in Irrigated Agriculture: Trends and Challenges in the Face of Emerging Demands* (Washington, D.C.: U.S. Dept. of Agriculture, Economic Research Service, 2012), 2.
4. "Imperial County Agriculture," Imperial County Farm Bureau, https://www.icfb.net/i-v-agriculture.
5. Barringer, 2011.
6. Tyreen L. Livingston, "Colorado Farmer's Influence on Flexible Water Policy Decisions: A Uniform Political Bloc?" (undergraduate thesis, University of Colorado Boulder, 2015).

Page 92

1. "Spain: Tornado Destroys 200 Hectares of Greenhouses," *FreshPlaza: Global Fresh Produce and Banana News*, January 8, 2018, http://www.freshplaza.com/article/187450/Spain-Tornado-destroys-200-hectares-of-greenhouses.
2. Birgita Dusse and Malte Meyer, "Social Europe!? Component for the Political Education of European Trade Unionists," Goethe University Frankfurt, https://www.uni-frankfurt.de/62312345/Social_Europe_English.pdf.
3. Ibid.
4. Felicity Lawrence, "Spain's Salad Growers Are Modern-Day Slaves, Say Charitles," *Guardian*, February 7, 2011.
5. "Food as Geopolitical Subjugation—Welcome to Plastic City, Almería!" *NaturPhilosophie*, April 29, 2015.

Page 94

1. "Pivot Irrigation in Saudi Arabia," NASA, May 23, 2013, https://svs.gsfc.nasa.gov/11290.
2. Ibid.
3. "Pivot Irrigation Systems Saudi Arabia," T-L Irrigation, accessed April 19, 2018, https://www.tlirr.com/blog/testimonial/jack-king/.
4. T-L Irrigation.
5. Frank Odenthal, "Sustainable Agriculture in the Desert," *fairplanet*, January 30, 2017, https://www.fairplanet.org/story/sustainable-agriculture-in-the-desert/.
6. Nathan Halverson, "What California Can Learn from Saudi Arabia's Water Mystery," *Reveal*, August 4, 2017, https://www.revealnews.org/article/what-california-can-learn-from-saudi-arabias-water-mystery/.
7. Ruth Schuster, "The Secret of Israel's Water Miracle and How It Can Help a Thirsty World," *Haaretz*, January 15, 2018, https://www.haaretz.com/science-and-health/how-israel-can-help-a-thirsty-world-1.5392651.
8. Odenthal, 2017.

Page 98

1. Harley Rustad, "*Big Lonely Doug,*" *The Walrus,* September 19, 2016, https://thewalrus.ca/big-lonely-doug/.

Page 100

1. "Indonesia's Fire and Haze Crisis," The World Bank, November 25, 2015, http://www.worldbank.org/en/news/feature/2015/12/01/indonesias-fire-and-haze-crisis.
2. Oliver Balch, "Indonesia's Forest Fires: Everything You Need to Know," *Guardian*, November 11, 2015, https://www.theguardian.com/sustainable-business/2015/nov/11/indonesia-forest-fires-explained-haze-palm-oil-timber-burning.
3. Ibid.
4. "Modern Oil Palm Cultivation," Food and Agriculture Organization of the United Nations, accessed April 3, 2018, http://www.fao.org/docrep/006/t0309e/T0309E01.htm.
5. "The African Oil Palm," Food and Agriculture Organization of the United Nations, accessed April 3, 2018, http://www.fao.org/docrep/004/ac126e/ac126e05.htm.
6. "Palm Oil & Forest Conversion," WWF, accessed April 3, 2018, http://wwf.panda.org/what_we_do/footprint/agriculture/about_palm_oil/environmental_impacts/forest_conversion/.
7. "Cutting Deforestation Out of Palm Oil," Greenpeace International, March 3, 2016, https://www.greenpeace.org/archive-international/en/publications/Campaign-reports/Forests-Reports/Cutting-Deforestation-Out-Of-Palm-Oil/.
8. "Borneo Deforested 30 Percent over Past 40 Years," *ScienceDaily*, July 16, 2014, accessed April 3, 2018, https://www.sciencedaily.com/releases/2014/07/140716141302.htm.
9. David L. A. Gaveau, Sean Sloan, Elis Molidena, Husna Yaen, Doug Sheil, Nicola K. Abram, Marc Ancrenaz, Robert Nasi, Marcela Quinones, Niels Wielaard, and Erik Meijaard, "Four Decades of Forest Persistence, Clearance and Logging on Borneo," *PLoS ONE* 9, no. 7 (2014).
10. "The Oil Palm," FAO Economic and Social Development Series: Better Farming Series, Institut africain pour le développement économique et social B.P. 8008, Abidjan, Côte d'Ivoire, 1990.
11. H. Zulkifli, M. Halimah, K. W. Chan, Y. M. Choo, and W. Mohd Basri, "Life Cycle Assessment for Oil Palm Fresh Fruit Bunch Production from Continued Land Use for Oil Palm Planted on Mineral Soil (part 2)," *Journal of Oil Palm Research* 22 (December 2010): 887–894.
12. "Malaysia: Economic Transformation Advances Oil Palm Industry," AOCS, accessed April 3, 2018, https://www.aocs.org/stay-informed/read-inform/featured-articles/malaysia-economic-transformation-advances-oil-palm-industry-september-2012.
13. Greenpeace International, 2016.
14. "Palm Oil Innovation Group," WWF, accessed April 4, 2018, http://wwf.panda.org/what_we_do/footprint/agriculture/about_palm_oil/solutions/palm_oil_innovation_group/.

Page 104

1. Ayansina Ayanlade and Nicolas Drake, "Forest Loss in Different Ecological Zones of the Niger Delta, Nigeria: Evidence from Remote Sensing," *GeoJournal* 81, no. 5 (2015): 717–735.

2. Glory O. Enaruvbe and Ozien P. Atafo, "Analysis of Deforestation Pattern in the Niger Delta Region of Nigeria," *Journal of Land Use Science* 11, no. 1 (2016): 113–130.

Page 106

1. Alan T. Bull, Juan A. Asenjo, Michael Goodfellow, and Benito Gómez-Silva, "The Atacama Desert: Technical Resources and the Growing Importance of Novel Microbial Diversity," *Annual Review of Microbiology* 70 (2016): 215–234.

2. "Countries with the Largest Lithium Reserves Worldwide 2017 (in Metric Tons)," Statistia, 2018, https://www.statista.com/statistics/268790/countries-with-the- largest-lithium-reserves-worldwide/.

3. "Lithium Mining in Chile," Lithium Mining—The Worldwide Website, accessed February 23, 2018, http://www.lithiummine.com/lithium-mining-in-chile.

4. Ibid.

5. Lilly G. Corenthal et al., "Regional Groundwater Flow and Accumulation of a Massive Evaporite Deposit at the Margin of the Chilean Altiplano," *Geophysical Research Letters* 43, no. 15 (2016): 8017–8025.

6. "The Lithium Test: How Chile Can Leverage Its Resource Advantage," Wharton University of Pennsylvania, accessed April 4, 2018, http://knowledge.wharton.upenn.edu/article/the-lithium-test-how-chile-can-leverage-its-resource-advantage/.

7. Dwight Bradley et al., "A Preliminary Deposit Model for Lithium Brines," USGS, https://oneworldlithium.com/wp-content/uploads/2017/12/OF13-1006-USGS-lithium-model.pdf.

8. Ibid.

9. Henry Sanderson, "Electric Car Demand Sparks Lithium Supply Fears," *Financial Times*, June 8, 2017.

10. Ibid.

11. Anna Hirtenstein, "Move over Tesla, Europe's Building Its Own Battery Gigafactories," *Bloomberg*, May 22, 2017.

Page 112

1. "Nigeria - Executive Summary," Export.gov, July 13, 2016, https://www.export.gov/apex/article2?id=Nigeria-Executive-Summary.

2. Adam Vaughan, "Oil in Nigeria: A History of Spills, Fines and Fights for Rights," *Guardian*, August 4, 2011, https://www.theguardian.com/environment/2011/aug/04/oil-nigeria-spills-fines-fights.

3. "Crude politics: A Broken Oil Industry is the Source of Many Woes," *The Economist*, March 28, 2015.

4. Adam Nossiter, "As Oil Thieves Bleed Nigeria, Report Says, Officials Profit," *New York Times*, September 19, 2013, https://www.nytimes.com/2013/09/20/world/africa/nigerias-politicians-profit-from-industrial-scale-oil-theft-report-says.html.

5. Nicole Crowder, "Deforestation vs. Daily Life: The Logging Industry in Nigeria Is Fueling Both," *Washington Post*, November 21, 2014, https://www.washingtonpost.com/news/in-sight/wp/2014/11/21/deforestation-vs-daily-life-the-logging-industry-in-nigeria-is-fueling-both/?utm_term=.a0d51d526783.

6. Glory O. Enaruvbe and Ozien P. Atafo, "Analysis of Deforestation Pattern in the Niger Delta Region of Nigeria," *Journal of Land Use Science* 11, no. 1 (2016): 113–130.

7. Matt Mossman, "How Nigeria's Oil Industry Went Local," *Foreign Affairs*, November 29, 2017.

8. Ibid.

Page 116

1. "Mineral Resource of the Month: Phosphate Rock," *Earth Magazine*, May 15, 2015, https://www.earthmagazine.org/article/mineral-resource-month-phosphate-rock.
2. *Environmental Aspects of Phosphate and Potash Mining*, International Fertilizer Industry Association and United Nations Environment Programme, 2001.
3. Ibid.
4. "Scientist Urges Government to Address 'Peak Phosphate' Risk," *Guardian*, July 14, 2010, https://www.theguardian.com/environment/2010/jul/14/oil-food.
5. "Mineral Resource of the Month: Phosphate Rock."
6. George H. S. Knibbs, *The Shadow of the World's Future: or, The Earth's Population Possibilities and the Consequences of the Present Rate of Increase of the Earth's Inhabitants* (New York: Arno Press, 1976).

Page 122

1. "Hambach Opencast Mine," RWE, accessed February 26, 2018, http://www.rwe.com/web/cms/en/60012/rwe-power-ag/fuels/hambach/.
2. "The World's Largest Diggers: In Pictures," *The Telegraph*, April 6, 2011, https://www.telegraph.co.uk/finance/newsbysector/industry/mining/8433249/The-worlds-largest-diggers-in-pictures.html?image=2.
3. "Hambach Opencast Mine," RWE.
4. "The World's Largest Diggers," *The Telegraph*.
5. As told to Jennifer Baichwal by a bagger operator at Hambach.
6. Katharina Wecker, "Environmental Activists Storm Open-pit Coal Mine Ahead of Climate Talks," *Deutsche Welle,* June 11, 2017, http://www.dw.com/en/environmental-activists-storm-open-pit-coal-mine-ahead-of-climate-talks/a-41248945.
7. *CO_2 from Fuel Emissions: Highlights* (Paris: International Energy Association Publications, 2017), https://www.iea.org/publications/freepublications/publication/CO_2EmissionsfromFuelCombustion-Highlights2017.pdf.

Page 124

1. Elian Hadj-Hamdi, "The Battle for Villages and Forests in Germany's Coal Country," Deutsche Welle, http://www.dw.com/en/the-battle-for-villages-and-forests-in-germanys-coal-country/a-39964913.
2. Ibid.

Page 128

1. "United States Remains the World's Top Producer of Petroleum and Natural Gas Hydrocarbons," Today in Energy, U.S. Energy Information Administration (EIA), June 7, 2017, https://www.eia.gov/todayinenergy/detail.php?id=31532.
2. Ibid.
3. "Multi-stage Fracking Sparked Energy Revolution," *CBC News*, April 1, 2013, http://www.cbc.ca/news/canada/calgary/multi-stage-fracking-sparked-energy-revolution-1.1338662.
4. Marin Katusa, "Don't Frack Me Up: Correcting Misinformation on Hydraulic Fracturing," *Forbes*, January 24, 2012.
5. Anjil Raval, "U.S. on Track to Become World's Largest Oil Producer," *Financial Times*, February 13, 2018.
6. Kelly Levin, "Carbon Dioxide Emissions from Fossil Fuels and Cement Reach Highest Point in Human History," World Resources Institute, November 22, 2013, http://www.wri.org/blog/2013/11/carbon-dioxide-emissions-fossil-fuels-and-cement-reach-highest-point-human-history.
7. "Graphic: Carbon Dioxide Hits New High," NASA, July 22, 2013, https://climate.nasa.gov/climate_resources/7/graphic-carbon-dioxide-hits-new-high/.
8. Levin, 2013.

Page 134

1. Terence Bell, "The World's 20 Largest Copper Mines," *The Balance*, November 3, 2017, https://www.thebalance.com/the-world-s-20-largest-copper-mines-2014-2339745.
2. David R. Fuller, "The Production of Copper in 6th Century Chile's Chuquicamata Mine," *JOM* 56, no. 11 (2004).
3. *Know Your Lifestyle: Introducing Sustainable Consumption in Second Chance Education* (Bonn: DVV International, 2014), http://www.knowyourlifestyle.eu/images/uploads/kyl_1_handy_english.pdf.
4. "Chuquicamata Copper Mine—Chile," Sisgeo, accessed June 8, 2018, https://www.sisgeo.com/uploads/schede/chuquicamata.pdf.
5. United States Geological Services statistics, 2009.
6. United States Environmental Protection Agency statistics.

Page 138

1. Madan M. Singh, *Water Consumption at Copper Mines in Arizona* (Phoenix: Arizona Department of Mines and Mineral Resources, 2010).
2. "Open Pit Mines, Southern Arizona: Image of the Day," NASA, February 8, 2010, https://earthobservatory.nasa.gov/IOTD/view.php?id=42555.
3. "TENORM: Copper Mining and Production Wastes," EPA, November 8, 2017, https://www.epa.gov/radiation/tenorm-copper-mining-and-production-wastes.
4. Singh, 2010.
5. Ibid.

Page 142

1. Andrew E. Kramer, "Russian City on Watch Against Being Sucked into the Earth," *New York Times*, April 10, 2012.

Page 146

1. Paul Veyret, Aubrey Diem, and Thomas M. Poulsen, "Alps," *Encyclopædia Britannica,* December 15, 2017, https://www.britannica.com/place/Alps.
2. Ibid.
3. "The Gotthard Base Tunnel," Federal Office of Transport (Swiss Confederation)—Swiss Federal Railways, April 2016, http://www.gottardo2016.ch/index-Dateien/SBB_Gottardo_Flyer_EN.pdf.
4. Ibid.

Page 148

1. V. Liguori, G. Rizzo, and M. Traverso, "Marble Quarrying: an Energy and Waste Intensive Activity in the Production of Building Materials," *WIT Transactions on Ecology and the Environment* 108 (2008): 197–207.
2. Ibid.
3. Ibid.
4. Ibid.
5. Ibid.

Page 154

1. *International Energy Outlook: 2017* (Washington, D.C.: U.S. Energy Information Administration, 2017), https://www.eia.gov/outlooks/ieo/pdf/0484(2017).pdf.
2. "Global Emissions," Center for Climate and Energy Solutions, January 4, 2018, https://www.c2es.org/content/international-emissions/.
3. "Wind Power Capacity Reaches 539 GW, 52.6 GW Added in 2017," World Wind Energy Association, February 12, 2018, http://www.wwindea.org/2017-statistics/.
4. "New Energy Outlook 2017," *Bloomberg New Energy Finance*, accessed March 9, 2018, https://about.bnef.com/new-energy-outlook/.

Page 156

1. Jonathan Watts, "Desert Tower Raises Chile's Solar Power Ambition to New Heights," *Guardian,* December 22, 2015, https://www.theguardian.com/ environment/2015/dec/22/desert-tower-raises-chiles-solar-power-ambition-to-new-heights.
2. Ibid.
3. Knvul Sheikh, "New Concentrating Solar Tower Is Worth Its Salt with 24/7 Power," *Scientific American,* July 14, 2016, https://www.scientificamerican.com/article/new-concentrating-solar-tower-is-worth-its-salt-with-24-7-power/.
4. "Solar Seeks Its Place under Spanish Sun," *Times LIVE,* March 28, 2018, https://www.timeslive.co.za/news/2018-03-28-solar-seeks-its-place-under-spanish-sun/.
5. "ES Executive Summary," *Renewables 2017 Global Status Report* (Paris: REN21 Secretariat, 2017), http://www.ren21.net/gsr-2017/pages/summary/summary/.
6. *International Energy Outlook: 2017* (Washington, D.C.: U.S. Energy Information Administration, 2017), https://www.eia.gov/outlooks/ieo/pdf/0484(2017).pdf.

Page 166

1. "BC's Coastal Biodiversity: The Highest in North America," Raincoast Conservation Foundation, October 5, 2011, http://www.raincoast.org/2011/05/bc-coastal-biodiversity/.
2. "History of Commercial Logging," British Columbia in a Global Context, June 12, 2014, https://opentextbc.ca/geography/chapter/7-3-history-of-commercial-logging/.
3. Ibid.
4. *British Columbia's Forests and Their Management* (British Columbia Ministry of Forests, 2003), https://www.for.gov.bc.ca/hfd/pubs/docs/mr/mr113/harvest.htm.
5. *British Columbia's Forest Industry and the B.C. Economy in 2016* (PricewaterhouseCoopers LLP, September 2017), https://www.co .org/wp-content/uploads/BC-Forest-Report-FINAL-Sept-2017.pdf.
6. "Vancouver Island Old-Growth Logging Speeding Up," Sierra Club B.C., March 6, 2018, https://sierraclub.bc.ca/vancouver-island-old-growth-logging-speeding-up/.
7. "Old-Growth Logging Rate Will Lead to Collapse on Vancouver Island," Sierra Club, July 16, 2017, https://sierraclub.bc.ca/old-growth-logging-collapse-vancouver-island/.
8. Gordon Hamilton, "B.C. Interior Lumber Supply Falling, Mills Threatened," *Times Colonist*, February 1, 2017, http://www.timescolonist.com/news/b-c/b-c-interior-lumber-supply-falling-mills-threatened-1.9711456.
9. Nelson Bennett, "B.C. Raw Log Exports to Asia Soar," *Business in Vancouver*, March 14, 2017.
10. Bob Williams, *Restoring Forestry in BC: The Story of the Industry's Decline and the Case for Regional Management* (Vancouver: Canadian Centre for Policy Alternatives, January 2018).
11. Andrew A. Duffy, "Raw-Log Exports Reach All-Time High: Research," *Times Colonist*, February 27, 2017.

Page 172

1. "Komodo National Park, Indonesia," WWF, http://wwf.panda.org/what_we_do/where_we_work/coraltriangle/coraltrianglefacts/places/komodonationalparkindonesia/.
2. "Komodo National Park," World Heritage Outlook, http://www.worldheritageoutlook.iucn.org/explore-sites/wdpaid/67725.
3. "The Facts on Great Barrier Reef Coral Mortality," Australia Government: Great Barrier Reef Marine Park Authority, March 6, 2016.
4. "Reef Health," Australia Government: Great Barrier Reef Marine Park Authority, June 29, 2017, http://www.gbrmpa.gov.au/about-the-reef/reef-health.
5. T. P. Hughes, J. T. Kerry, M. Álvarez-Noriega, J. G. Álvarez-Romero, K. D. Anderson, A. H. Baird, R. C. Babcock, et al., "Global Warming and Recurrent Mass Bleaching of Corals," *Nature* 543, no. 7645 (2017): 373–377.
6. Damien Cave and Justin Gillis, "Large Sections of Australia's Great Reef Are Now Dead, Scientists Find," *New York Times*, March 15, 2017.

Page 178

1. "Rhino Horn Use: Fact vs. Fiction," *PBS Nature,* August 20, 2010, http://www.pbs.org/wnet/nature/rhinoceros-rhino-horn-use-fact-vs-fiction/1178/.
2. Ibid.
3. Ol Pejeta Conservancy, "This morning, we gave Sudan, the last male northern white rhino, a fitting tribute in honour of his life and his great work as a rhino conservation ambassador," Facebook, March 31, 2018, https://www.facebook.com/OlPejetaConservancy/posts/10160185183990324.

Page 180

1. *Living Planet Report 2016: Risk and Resilience in a New Era,* (Gland: World Wildlife Fund, 2016).
2. Patrick Omondi and Shadrack Ngene, "The National Elephant Conservation and Management Strategy (2012–2021) at a Glance," *The George Wright Forum* 29, no. 1 (2012): 90–92.
3. "The Great Elephant Census: A Paul G. Allen Project Country-by-Country Findings," The Great Elephant Census, https://static1.squarespace.com/static/5304f39be4b0c1e749b456be/t/57c71f5fcd0f68b39c3f4bfa/1472667487326/GEC+Results+Country+by+Country+Findings+Fact+Sheet_FINAL_8+26+2016.pdf.
4. Scott Hitch, "Losing the Elephant Wars: CITES and the 'Ivory Ban,'" *Georgia Journal of International Comparative Law* 27, no. 167 (1998).
5. Damian Zane, "Kenya's Ivory Inferno: Does Burning Elephant Tusks Destroy Them?" *BBC News*, April 29, 2016, http://www.bbc.com/news/world-africa-34313745.
6. Ibid.
7. Jeffrey Gettleman, "Kenya Burns Elephant Ivory Worth $105 Million to Defy Poachers," *New York Times*, April 30, 2016, https://www.nytimes.com/2016/05/01/world/africa/kenya-burns-poached-elephant-ivory-uhuru-kenyatta.html.
8. Ibid.
9. "Press and Media," Save the Elephants, March 29, 2017, http://www.savetheelephants.org/about-ste/press-media/?detail=dramatic-changes-in-china-s-ivory-trade.
10. Rachel Bale, "China Shuts Down Its Legal Ivory Trade," *National Geographic*, December 30, 2017, https://news.nationalgeographic.com/2017/12/wildlife-watch-china-ivory-ban-goes-into-effect/.

Page 186

1. Michael J. Chase, Scott Schlossberg, Curtice R. Griffin, Philippe J. C. Bouché, Sintayehu W. Djene, Paul W. Elkan, Sam Ferreira, et al., "Continent-Wide Survey Reveals Massive Decline in African Savannah Elephants," *PeerJ* 4, no. e2354 (2016).
2. Ibid.

List of Works

[A] = AGO only
[N] = NGC only
[AN] = AGO + NGC

Photographs (148.6 × 198.1 cm)

Edward Burtynsky
Pigment inkjet prints
Courtesy of the artist and
Nicholas Metivier Gallery, Toronto
© Edward Burtynsky

Basque Coast #1, UNESCO Geopark, Zumaia, Spain, 2015 [A]
Page 61

Brine Wells #1, Salt Flats, Atacama Desert, Chile, 2017 [A]
Page 108

Burned Ivory Tusks #1, May 1, Nairobi, Kenya, 2016 [A]
Page 181

Burned Ivory Tusks #2, May 1, Nairobi Kenya, 2016 [N]
Page 220

Cerro Dominador Solar Project #1, Atacama Desert, Chile, 2017 [AN]
Page 157

Chino Mine #5, Silver City, New Mexico, USA, 2012 [N]
Page 141

Chuquicamata Copper Concentrator and Smelter, Calama, Chile, 2017 [N]
Page 137

Chuquicamata Copper Mine Overburden #1, Calama, Chile, 2017 [A]
Page 135

Clearcut #1, Palm Oil Plantation, Borneo, Malaysia, 2016 [AN]
Page 101

Clearcut #3, Palm Oil Plantation, Borneo, Malaysia, 2016 [N]
Page 103

Clearcut #5, Vancouver Island, British Columbia, Canada, 2017 [A]
Page 96

Coal Mine #1, North Rhine, Westphalia, Germany, 2015 [AN]
Page 123

Coal Mining, Near Gillette, Wyoming, USA, 2015 [A]
Page 132

Dandora Landfill #1, Nairobi, Kenya, 2016 [N]
Page 65

Dandora Landfill #3, Plastics Recycling, Nairobi, Kenya, 2016 [AN]
Page 63

Dryland Farming #40, Monegrillo, Spain, 2010 [A]
Page 87

Eko Atlantic Development #1, Lagos, Nigeria, 2016 [A]
Page 71

Flood Damaged Cars, Royal Purple Raceway, Baytown, Texas, USA, 2017 [A]
Page 83

Fracking, Near Gillette, Wyoming, USA, 2015 [AN]
Page 129

Freeman Island, Long Beach, California, USA, 2017 [A]
Page 163

Greenhouses #2, El Ejido, Southern Spain, 2010 [A]
Page 93

Highway #8, Santa Ana Freeway, Los Angeles, California, USA, 2017 [AN]
Page 81

Imperial Valley #4, California, USA, 2009 [A]
Page 89

Imperial Valley #5, Holtville, California, USA, 2009 [N]
Page 91

Lithium Mines #1, Salt Flats, Atacama Desert, Chile, 2017 [AN]
Page 107

Makoko #2, Lagos, Nigeria, 2016 [A]
Page 73

Morenci Mine #1, Clifton, Arizona, USA, 2012 [N]
Page 139

Morenci Mine #2, Clifton, Arizona, USA, 2012 [AN]
Page 140

Oil Bunkering #1, Niger Delta, Nigeria, 2016 [AN]
Page 113

Oil Bunkering #4, Niger Delta, Nigeria, 2016 [AN]
Page 114

Oil Refineries and Storage, Carson, California, USA, 2017 [A]
Page 165

Petrochemical Plants, Baytown, Texas, USA, 2017 [AN]
Page 159

Phosphor Tailings #5, Near Lakeland, Florida, USA, 2012 [A]
Page 119

Phosphor Tailings #6, Near Lakeland, Florida, USA, 2012 [A]
Page 115

Phosphor Tailings Pond #4, Near Lakeland, Florida, USA, 2012 [A]
Page 117

PS10 Solar Power Plant, Seville, Spain, 2013 [A]
Page 158

Salinas #5, Aquaculture, Cádiz, Spain, 2013 [N]
Page 85

Salt Pan #18, Little Rann of Kutch, Gujarat, India, 2016 [A]
Page 111

Satellite Capture, Near Buraydah, Saudi Arabia, 2018 [A]
Page 189

Saw Mills #1, Lagos, Nigeria, 2016 [AN]
Page 105

South Bay Pumping Plant #1, Near Livermore, California, USA, 2009 [A]
Page 155

Tetrapods #1, Dongying, China, 2016 [AN]
Page 69

Tyrone Mine #3, Silver City, New Mexico, USA, 2012 [A]
Page 133

Uralkali Potash Mine #2, Berezniki, Russia, 2017 [N]
Page 143

Uralkali Potash Mine #4, Berezniki, Russia, 2017 [A]
Page 144

Photographs (118.1 × 158.8 cm)

Edward Burtynsky
Pigment inkjet prints
Courtesy of the artist and Nicholas Metivier Gallery, Toronto
© Edward Burtynsky

Basque Coast #2, UNESCO Geopark, Zumaia, Spain, 2015 [N]
Page 197

Brine Wells #2, Salt Flats, Atacama Desert, Chile, 2017 [N]
Page 109

Clearcut #3, Palm Oil Plantation, Borneo, Malaysia, 2016 [N]
Page 103

Dryland Farming #40, Monegrillo, Spain, 2010 [N]
Page 87

Greenhouses #2, El Ejido, Southern Spain, 2010 [N]
Page 92

Log Booms #1, Vancouver Island, British Columbia, Canada, 2016 [N]
Page 97

Phosphor Tailings Pond #4, Near Lakeland, Florida, USA, 2012 [N]
Page 117

Salt Pan #21, Little Rann of Kutch, Gujarat, India, 2016 [N]
Page 110

South Bay Pumping Plant #1, Near Livermore, California, USA, 2009 [N]
Page 155

Uralkali Potash Mine #6, Berezniki, Russia, 2017 [N]
Page 145

Films

Jennifer Baichwal and Nicholas de Pencier
HD video loops
Courtesy of the artists

Bagger 291, Hambach Lignite Mine, Germany, 2018 [AN]
Page 125

Coal Trains, Wyoming, USA, 2018 [AN]
Page 131

Coral Bleaching, Great Barrier Reef, Australia, 2018 [AN]
Page 225

Dandora Landfill, Nairobi, Kenya, 2018 [AN]
Page 67

Elephant Tusk Burn, Nairobi National Park, Kenya, 2018 [AN]
Page 187

Exploding Danger Trees #2, Cathedral Grove, British Columbia, Canada, 2018 [AN]
Page 204

Felling Danger Trees, Cathedral Grove, British Columbia, 2018 [AN]

Gotthard Base Tunnel, Gotthard, Switzerland, 2018 [AN]
Page 147

Hambach Lignite Mine, Germany, 2018 [AN]
Page 127

Norilsk, Russia, 2018 [AN]

Oil Refineries, Texas, USA, 2018 [AN]
Page 161

Phosphate Mines, Florida, USA, 2018 [AN]
Page 121

Murals and Film Extensions

Edward Burtynsky
Pigment inkjet prints on adhesive vinyl
Courtesy of the artist and
Nicholas Metivier Gallery, Toronto
© Edward Burtynsky

Jennifer Baichwal and
Nicholas de Pencier
HD video
Courtesy of the artists

Carrara Marble Quarries, Cava di Canalgrande #2, Carrara, Italy, 2016 [A]
304.8 × 609.6 cm
Page 151

Carving Studio, Carrara, Italy, 2018
Page 153

Cava di Canalgrande, Carrara, Italy, 2018
Page 152

Falling Slabs, Carrara, Italy, 2018
Page 152

Cathedral Grove #1, Vancouver Island, British Columbia, Canada, 2017–2018 [AN]
243.8 × 731.5 cm
Page 169

Clearcuts, Vancouver Island, Canada, 2018
Page 170

Tree Felling, Vancouver Island, Canada, 2018
Page 171

Old-Growth Tree Portraits, Vancouver Island, Canada, 2018
Page 171

Mushin Market Intersection, Lagos, Nigeria, 2016–2018 [AN]
365.8 × 653.8 cm
Page 77

Makoko, Lagos, Nigeria, 2018
Page 78

Market Walks, Lagos, Nigeria, 2018
Page 78

Redeemed Christian Church of God, Lagos, Nigeria, 2018
Page 79

Pengah Wall #1, Komodo National Park, Indonesia, 2017–2018 [AN]
304.8 × 609.6 cm
Page 175

Bleached Coral, Great Barrier Reef, Australia, 2018
Page 176

Panoply of Reef Life, Komodo National Park, Indonesia, 2018
Page 177

Reef Portraits, Komodo National Park, Indonesia, and Great Barrier Reef, Australia, 2018
Page 177

Augmented Reality

Edward Burtynsky, Jennifer Baichwal, and Nicholas de Pencier
Photogrammetric Augmented Reality (AR)
Courtesy of the artists

AR #2, President Kenyatta's Tusk Pile, April 28, Nairobi, Kenya, 2016 [AN]
Page 185

AR #3, Big Lonely Doug, Vancouver Island, British Columbia, Canada, 2016 [AN]
Page 99

AR #4, Sudan, The Last Male Northern White Rhinoceros, Nanyuki, Kenya, 2016 [AN]
Page 179

The Art Gallery of Ontario is partially funded by the Ontario Ministry of Culture. Additional operating support is received from the City of Toronto, the Department of Canadian Heritage, and the Canada Council for the Arts.

Contemporary programming at the Art Gallery of Ontario is supported by

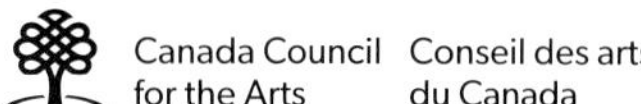

Goose Lane Editions acknowledges the generous support of the Government of Canada, the Canada Council for the Arts, and the Government of New Brunswick.

Printed and bound in Canada
10 9 8 7 6 5 4 3 2

Library and Archives Canada Cataloguing in Publication

Anthropocene : Burtynsky, Baichwal, de Pencier.

Co-published by: Goose Lane Editions.

Catalogue of an exhibition organized by the Art Gallery of Ontario and the Canadian Photography Institute of the National Gallery of Canada, in partnership with Fondazione MAST.

Edited by Sophie Hackett, Andrea Kunard and Urs Stahel.

ISBN 978-1-988788-04-3 (hardcover).—
ISBN 978-1-77310-097-5 (Goose Lane Editions)

1. Human ecology in art—Exhibitions.
2. Nature—Effect of human beings on—Exhibitions.
3. Global environmental change—Exhibitions.
4. Arts, Canadian—21st century—Exhibitions.
I. Kunard, Andrea, editor
II. Stahel, Urs, editor
III. Hackett, Sophie, 1971–, editor
IV. Art Gallery of Ontario, organizer, host institution
V. Canadian Photography Institute, organizer
VI. Fondazione MAST, organizer
VII. Burtynsky, Edward, 1955–. Works. Selections.
VIII. Baichwal, Jennifer. Works. Selections.
IX. De Pencier, Nick. Works. Selections.

N8217.E28A58 2018 709.71 C2018-904045-9

Published in conjunction with the exhibition
Anthropocene

Art Gallery of Ontario
Toronto, Ontario, Canada
September 28, 2018–January 6, 2019

National Gallery of Canada
Ottawa, Ontario, Canada
September 28, 2018–February 24, 2019

Fondazione MAST
Bologna, Italy
Spring 2019

Anthropocene is organized by the Art Gallery of Ontario and the Canadian Photography Institute of the National Gallery of Canada, in partnership with Fondazione MAST.

Art Gallery of Ontario
317 Dundas Street West
Toronto, Ontario M5T 1G4
Canada
www.ago.ca

Goose Lane Editions
500 Beaverbrook Court, Suite 330
Fredericton, New Brunswick E3B 5X4
Canada
www.gooselane.com

IN COLLABORATION WITH
National Gallery of Canada
380 Sussex Drive
Ottawa, Ontario K1N 9N4
Canada
www.gallery.ca

Fondazione MAST
Via Speranza, 42
40133 Bologna, Italy
www.mast.org

ART GALLERY OF ONTARIO AND NATIONAL GALLERY OF CANADA

PRESENTING SPONSOR
Scotiabank

IN PARTNERSHIP WITH
TELUS

ART GALLERY OF ONTARIO

LEAD SUPPORTER
Hal Jackman Foundation

GENEROUS SUPPORT FROM
Greg & Susan Guichon
Richard M. Ivey and Richard & Donna Ivey
Suzanne Ivey Cook and Rosamond Ivey
Robin & David Young

GENEROUS ASSISTANCE FROM
Michael Barnstijn & Louise MacCallum
The McLean Foundation
Gretchen & Donald Ross

WITH ADDITIONAL ASSISTANCE FROM
Donner Canadian Foundation

GOVERNMENT PARTNERS
Canada Council for the Arts
Ontario Cultural Attractions Fund

CURATORIAL TEAM

Sophie Hackett
Curator of Photography
Art Gallery of Ontario

Andrea Kunard
Associate Curator
Canadian Photography Institute of the National Gallery of Canada

Urs Stahel
Curator
Fondazione MAST

PUBLICATION

MANAGING EDITOR
Jim Shedden

PRODUCTION AND COPY EDITORS
Amy Lam
Sarah Liss

FRENCH AND ITALIAN PRODUCTION EDITORS
Laurel Saint-Pierre
Marina Rotondo, Fondazione MAST

PUBLISHING COORDINATOR
Robyn Lew

FRENCH TRANSLATORS
Dominique Denis
Marine Van Hoof

ITALIAN TRANSLATORS
Arianna Dagnino, PhD, and Stefano Gulmanelli, PhD
Giulia De Gasperi, PhD
Valentina Tortelli, NTL Traduzione, Firenze

GERMAN TRANSLATOR
Ishbel Flett

RESEARCHERS
Cathleen Evans
Kate Pocock
Zeesy Powers

FACT-CHECKERS
Megan Jones
Lucy Uprichard

PROOFREADER
Judy Phillips

FRENCH PROOFREADER
Re:word Communications

ITALIAN PROOFREADER
Sebastiano Bazzichetto

DESIGN
The Office of Gilbert Li

PRE-PRESS AND PRINTING
Type A Print Inc.

AGO EXHIBITION

CURATOR
Sophie Hackett

CHIEF CURATOR
Julian Cox

PROJECT MANAGER
Hillary Taylor

INTERPRETATION AND ENGAGEMENT
Nadia Abraham
Shiralee Hudson Hill
Melissa Smith

CURATORIAL ADMINISTRATIVE ASSISTANT
Jill Offenbeck

EDITORS
Amy Lam
Sarah Liss

DESIGN AND GRAPHICS
Katy Chey
Aleksandra Grzywaczewska
Malene Hjørngaard
Kristina Ljubanovic

Exhibitions and Collections

CHIEF, EXHIBITIONS AND COLLECTIONS
Christy Thompson

DIRECTOR OF EXHIBITIONS
Jessica Bright

REGISTRATION
Donna Austria
Cindy Brouse
Jerry Drozdowsky
Tim Hardacre
Joel Herman
Dale Mahar
Curtis Strilchuk

COLLECTION INFORMATION
Tracy Mallon-Jensen
Liana Radvak
Olga Zotova

CONSERVATORS
Cyntia Karnes
Sherry Phillips
Katharine Whitman
Sjoukje Van der Laan

LOGISTICS AND ART SERVICES
Curtis Amisich
Michael Beynon
Andrew Bugden
Scott Cameron
Corinne Carlson
Marco Cheuk
Patric Colosimo
Brian Davis
Mark Dudiak
Randal Fedje
Tina Giovinazzo
Eric Glavin
Brian Groombridge
Roland Hardy
Iain Hoadley
Matthew Janisse
Ruth Jones
Charles Kettle
David Kinsman
Jason Laudadio
Alison Lindsay
Paul Mathiesen
Doug Moore
Ben Oakley
Hagop Ohannessian
Jacques Oule
Angelo Pedari
Sabine Schaefer
Carmen Schroeder
Damian Seguin
Harry Singh
Jelena Sisko
David Stasyna
Craig Whiteside
Darin Yorston
Tanya Zhilinsky

Public Programming and Learning

ACTING HEAD, PUBLIC PROGRAMMING & LEARNING
Keri Ryan

DIRECTOR, PUBLIC PROGRAMS
Devyani Saltzman

PUBLIC PROGRAMMING
Kathleen McLean

MEDIA PRODUCTION
Matthew Scott

NGC EXHIBITION

The National Gallery of Canada extends its gratitude to all staff who assisted in this project.

MAST EXHIBITION

Fondazione MAST wishes to thank its staff and all those who contributed to make the exhibition possible in Bologna.

ANTHROPOCENE ARTIST ACKNOWLEDGEMENTS

SUPPORTERS OF *THE ANTHROPOCENE PROJECT*

Michael Barnstijn
Timothy Dattels
Kristine Johnson
Louise MacCallum
Roberto and Mary Ellen Marziale
Scott Mead
Bryan Pearson
Sally Peterson
Jovanka and Hans Porsche
Dr. Mark Pruzanski
Heather Reisman
Dr. Randolph Staab

THE ANTHROPOCENE PROJECT ADVISORY GROUP

Margaret Atwood
Dominic Barton
Stewart Brand
Robert Carsen
Graeme Gibson
Vicki Heyman
Victoria Jackman
Elizabeth Kolbert
Nicholas Metivier
Paul Roth

RESEARCH ASSOCIATES

Ashley Brook
Cathleen Evans
Joseph Hartman
Carlie Macfie
Jim Panou
Zeesy Powers
Marcus Schubert

COPY EDITORS

Cathleen Evans
Kate Pocock

ASSOCIATE COPY EDITORS

Jennifer Baichwal
Kyra Church
Carlie Macfie
Karen Machtinger

BURTYNSKY STUDIO

Joseph Hartman
Connie Hitzeroth
Julia Johnston
Karen Machtinger
Jim Panou
Laura Margaret Ramsey
Mike Reid
Marcus Schubert
Paul Sergeant
Alanna Smith

MERCURY FILMS

Yvonne Bayer
Andrew Beach
Ashley Brook
Helen Bucknell
Cathleen Evans
Carlie Macfie
Chris Niesing
Roland Schlimme
David Schmidt
Nadia Tavazzani

TORONTO IMAGE WORKS

Jeannie Baxter
Daniel Ebert
Francine Lemieux

SPECIAL THANKS

Elvina Baichwal
Mary Buratynsky
Katya and Meagan Burtynsky
Honor and Michael de Pencier
Magnus and Anna de Pencier
Ollie Moffatt, Krys Semmens, and Michael Burtynsky

GALLERY REPRESENTATION FOR EDWARD BURTYNSKY

Bryce Wolkowitz, New York
Enrica Viganò, Admira, Milan
Galerie Springer, Berlin
Howard Greenberg, New York
Matthew Flowers, London
Nicholas Metivier, Toronto
Paul Kuhn, Calgary
Robert Koch, San Francisco
Sundaram Tagore, Hong Kong / Singapore
Weinstein Hammons, Minneapolis

The artists wish to extend our profound appreciation to the talented and committed collaborators, colleagues, and patrons who have made the realization of *The Anthropocene Project* possible.

To represent the global scope of the Anthropocene, we took our cameras to each of our planet's continents (save Antarctica), often in challenging work environments. We were accompanied on most of these expeditions by Jim Panou and Mike Reid. Their expertise has elevated the work enormously.

The Anthropocene Project has been significantly shaped and strengthened by the curatorial team of Sophie Hackett, Andrea Kunard, and Urs Stahel. Without the initial faith put in us by Stephan Jost at the Art Gallery of Ontario, Marc Mayer at the National Gallery of Canada, and Isabella Seràgnoli and Mariella Criscuolo at MAST, we might never have attempted a project of this scale or complexity.

We extend both appreciation and respect to the members of the Anthropocene Working Group—especially Jan Zalasiewicz and Colin Waters, who have offered their wisdom, insight, and knowledge on a number of fronts.

Finally, deep gratitude goes to our teams at Mercury Films, the Burtynsky Studio, and to Julia Johnston, the Director of Partnerships for *The Anthropocene Project,* all of whom have put attentive, tireless effort and dedication towards this project's realization.

Pages 251–252:
At work 1,000 feet below the earth's surface in the Potash Mines of Berezniki, Siberia, 2017.

Pages 254–255:
Nicholas de Pencier and Jennifer Baichwal on location in Carrara, Italy. Photo courtesy of Anthropocene Films Inc. © 2018.

Page 256:
Jennifer Baichwal, Nicholas de Pencier, and Edward Burtynsky working in Northern British Columbia, Canada, 2012.

BACK COVER:
Edward Burtynsky, *Mushin Market Intersection, Lagos, Nigeria* (detail), 2016.

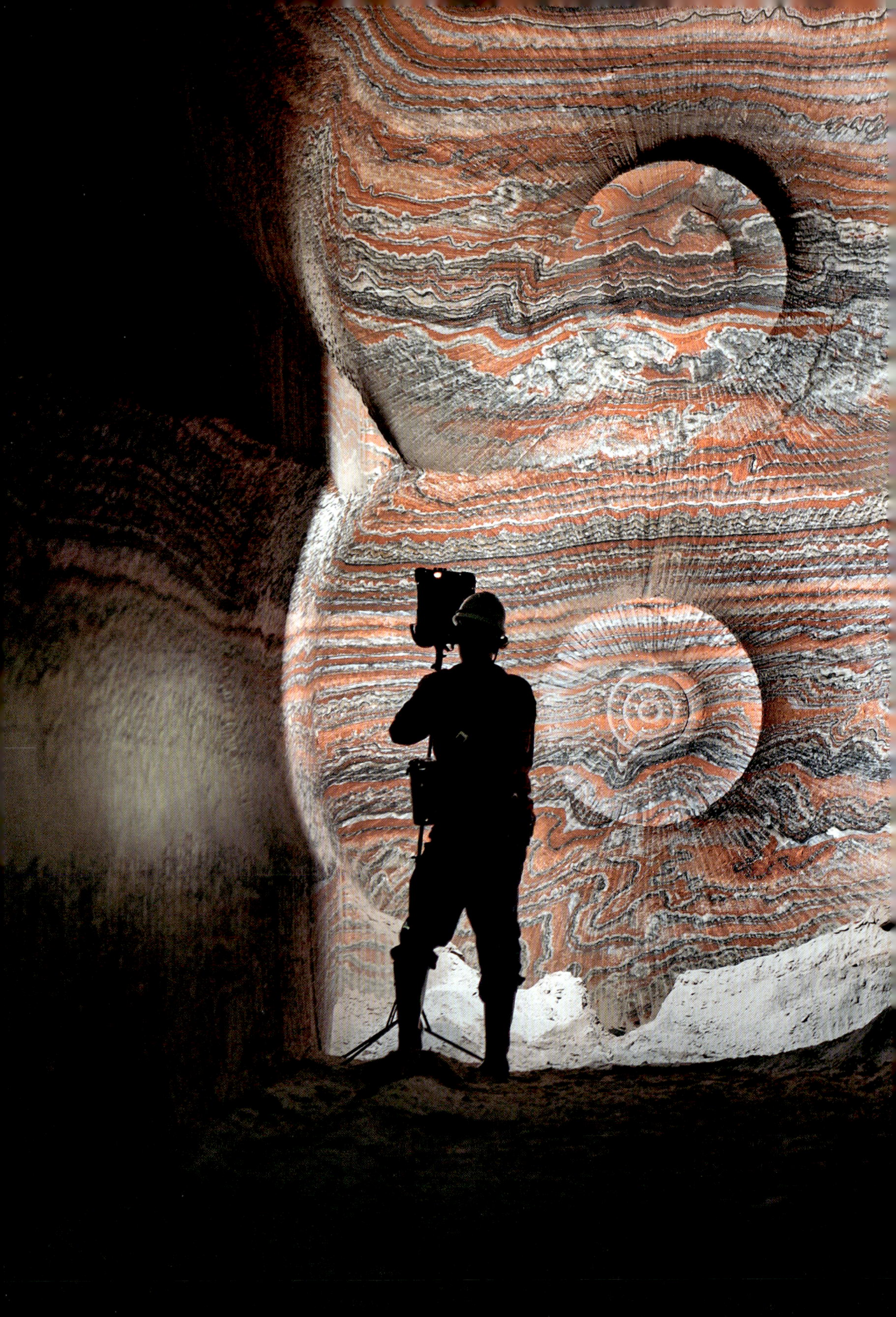

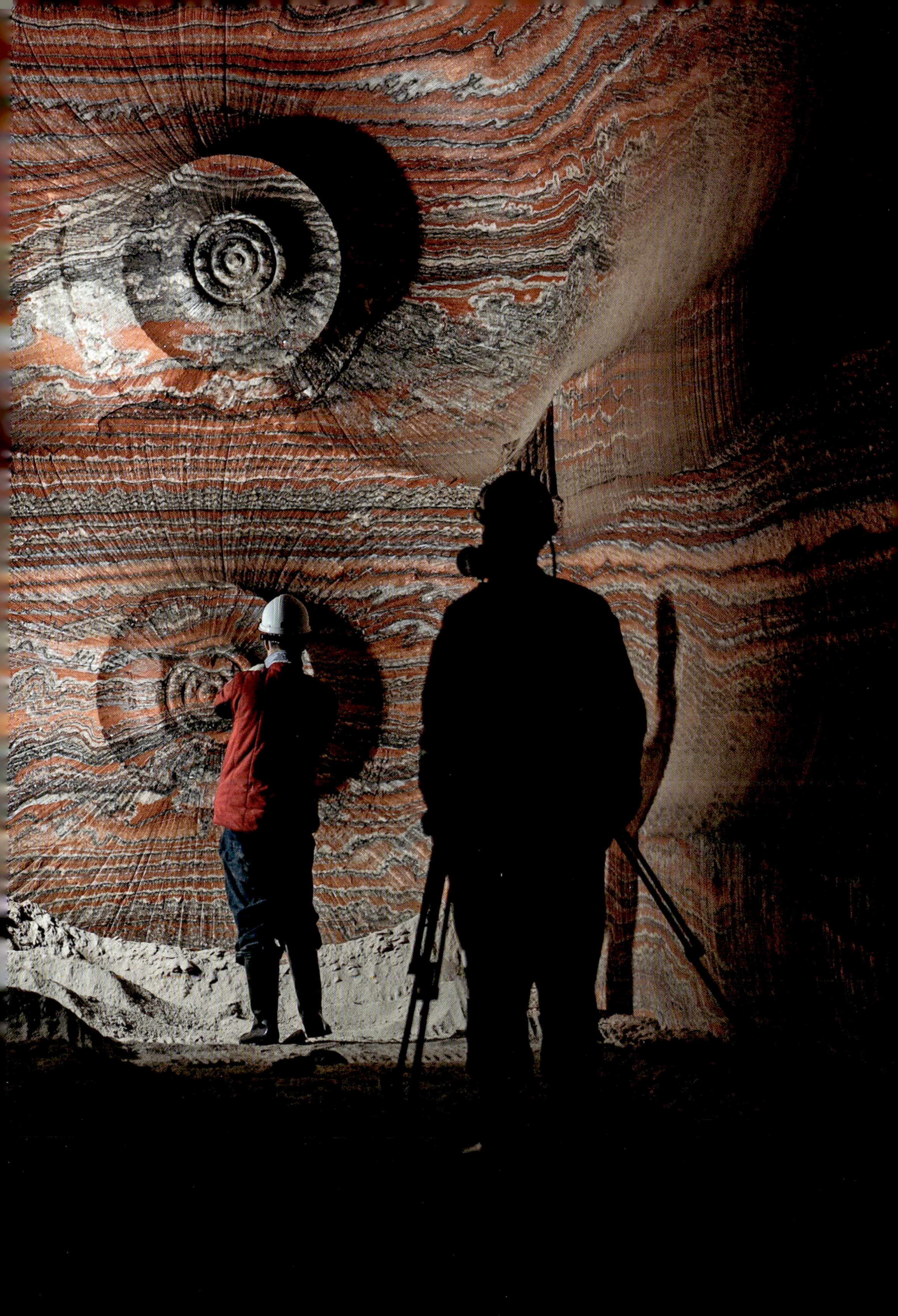

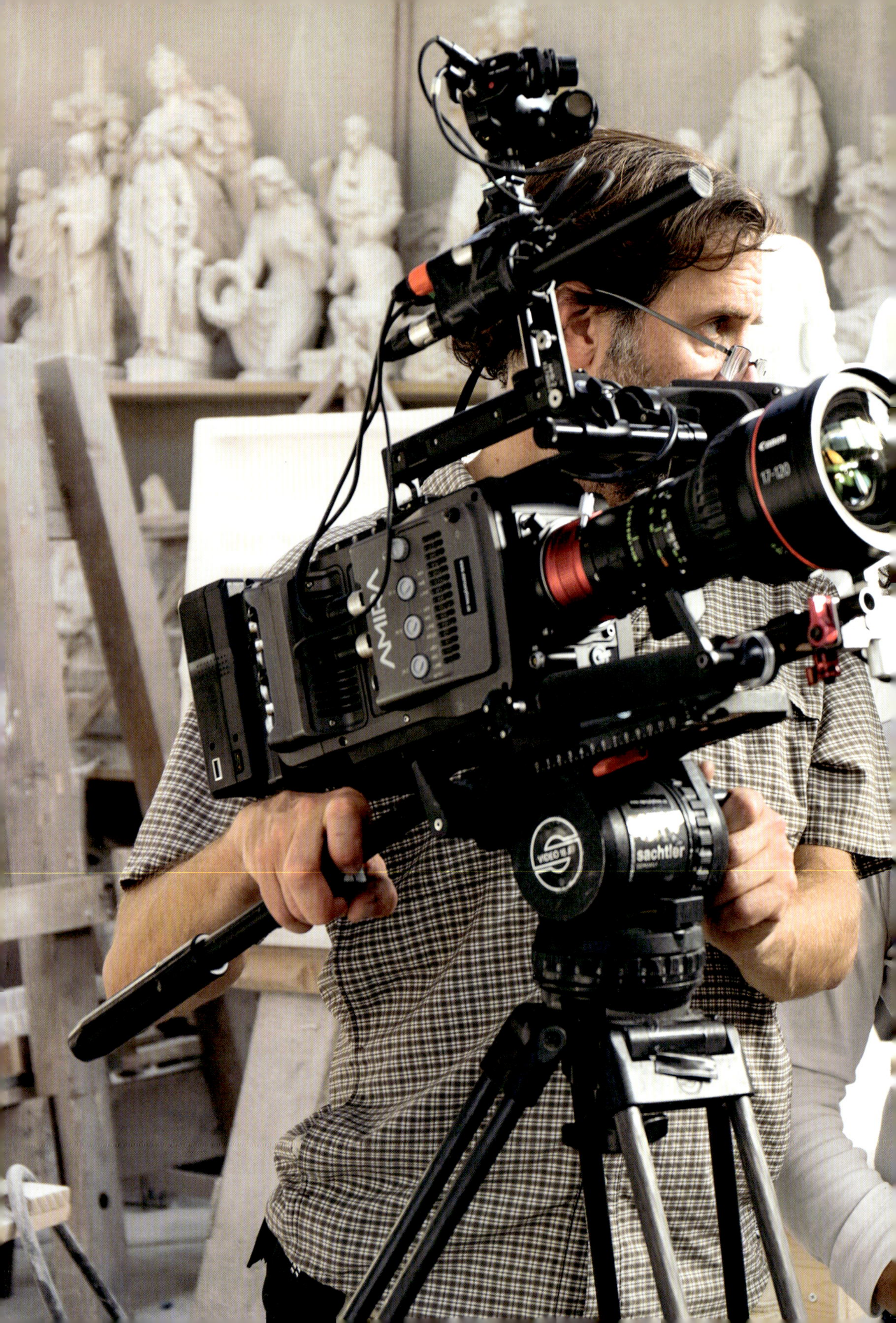
AMIRA
Canon
17-120
sachtler

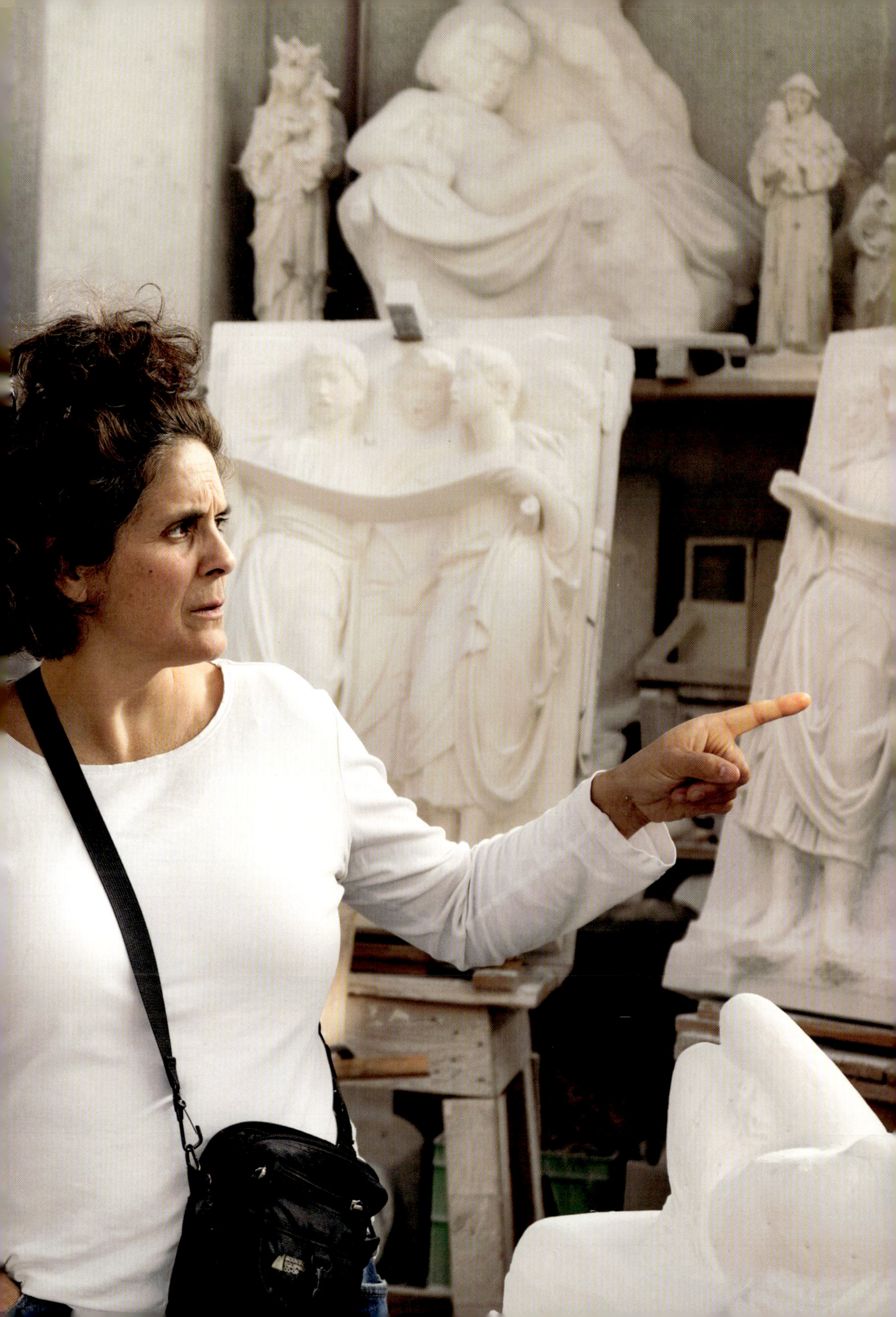